T0301368

The Global Challenge of Encouraging Sustainable Living

For Beth, Denise and Jamie
Three women who sustain all those around them – Steve

For Mum
Thanks for your love and continued support – Shane

For Nanna Bertha
With thanks for all the happy memories – Michael

The Global Challenge of Encouraging Sustainable Living

Opportunities, Barriers, Policy and Practice

Edited by

Shane Fudge

Lecturer in Energy Policy, University of Exeter, UK

Michael Peters

Lecturer in Energy Policy, University of Reading, UK

Steven M. Hoffman

Professor, University of St. Thomas, St. Paul, Minnesota, USA

Walter Wehrmeyer

Reader, University of Surrey, UK

Edward Elgar

Cheltenham, UK • Northampton, MA, USA

Published by
Edward Elgar Publishing Limited
The Lypiatts
15 Lansdown Road
Cheltenham
Glos GL50 2JA
UK

Edward Elgar Publishing, Inc.
William Pratt House
9 Dewey Court
Northampton
Massachusetts 01060
USA

A catalogue record for this book
is available from the British Library

Library of Congress Control Number: 2013942226

This book is available electronically in the ElgarOnline.com
Economics Subject Collection, E-ISBN 978 1 78100 375 6

ISBN 978 1 78100 374 9

Typeset by Servis Filmsetting Ltd, Stockport, Cheshire
Printed and bound in Great Britain by T.J. International Ltd, Padstow

Contents

Contributors

Wokje Abrahamse currently works as a Banting postdoctoral fellow in the Department of Psychology at the University of Victoria in Canada. She obtained her PhD in social and behavioural sciences at the University of Groningen, the Netherlands. Her research focuses on human behaviour in relation to environmental issues. She studies the effectiveness of interventions (such as information provision) to encourage the adoption of environmentally friendly behaviours. Wokje also examines factors related to engagement in environmentally friendly behaviours, such as attitudes, social norms and values.

Colin Ashton-Graham is a behavioural economist with 20 years experience in the development, delivery and evaluation of interventions to effect sustainable behaviour change. He has applied these skills to infrastructure and service planning in the transport sector and to demand management of water, energy, waste, garden nutrients (river health), physical activity (population health) and transport. He is now an independent consultant offering his skills and experience to government, education and not-for-profit clients with a passion for facilitating more sustainable community outcomes. Ashton-Graham is currently the technical lead on behaviour change programmes for the Water Corporation, Department of Sport and Recreation, Swan River Trust and several Local Government Authorities in Australia.

Professer Subhes C. Bhattacharyya is an internationally renowned energy specialist with more than 25 years of energy sector expertise in various capacities. He has extensively written on energy issues and brings a multidisciplinary background covering engineering, economics, regulatory skills and project management. He focuses on energy issues in developing countries and has worked extensively in South Asia, South East Asia and sub-Saharan Africa. Bhattacharyya is the author of *Energy Economics: Concepts, Issues, Markets and Governance* (2011) and has edited *Rural Electrification through Decentralised Off-grid System in Developing Countries* (2013). He is Professor of Energy Economics and Policy at the Institute of Energy and Sustainable Development, De Montfort University, Leicester, UK.

Mathieu Brugidou is Senior Researcher at EDF Research and Development, and was Associate Senior Researcher at PACTE, a CNRS social science research unit. He received a PhD and habilitation (HDR) in political science from Sorbonne-Paris-I University, France in 1992. His research interests include analysis of political discourse, public opinion and debate about energy and methods of analysis of textual data applied in social science (non-directive interview and open-ended question). Brugidou is a trainer for adult classes in different Schools (Sciences Po, Université Paris I Sorbonne). His books include *L'opinion et ses publics* (2008), *Le débat public: un risque démocratique* (2009, with D. Boy), *Le Grenelle de l'environnement: Acteurs, discours, effets* (2012, with D. Boy, C. Halpern and P. Lascoumes.

Roland Clift studied chemical engineering at Trinity College, Cambridge, UK achieving first class honours in 1964. He received a PhD from McGill University in 1970 for work on particle-fluid interactions, which became his main research area (at McGill and Cambridge, then Surrey University) in subsequent years. He became Head of the Department of Chemical Engineering at the University of Surrey in 1981 and a growing interest in the application of engineering principles to environmental issues led him to establish the Centre for Environmental Strategy (CES), a multidisciplinary group of engineers, scientists and social scientists, there in 1992. In this centre he was an advocate of clean technology, life cycle assessment and sustainable development. Clift is presently Professor Emeritus at the CES. He is a Fellow of the Institution of Chemical Engineers, the Royal Academy of Engineering and the Royal Society and in 1994 was made an OBE for his initiative in promoting research, and a CBE in 2006 for services to the environment. He received the Sir Frank Whittle Medal of the Royal Academy of Engineering for 2003, for 'an outstanding and sustained engineering achievement contributing to the well-being of the nation'. He received the 2007 Hanson Medal of the Institution of Chemical Engineers.

Judith de Groot is Senior Lecturer in Marketing at the University of Bath, UK. Her research interests include explaining and changing pro-environmental behaviours. De Groot's main focus includes how values and norms affect pro-environmental behaviour and how we can frame messages to activate values and norms thereby promoting pro-environmental behaviours.

Sophie Emmert studied public administration at the University of Twente, the Netherlands and works as a researcher and consultant at the TNO Institute of Environment and Geosciences. Her current research activities and interests focus on system innovations, transitions and transition management. As a co-project manager she worked (2005–08) on the European Union (EU) funded SCORE! Project, which aimed to initiate societal system innovations on Sustainable Consumption and Production together with different societal actors. Emmert also works in SCOPE2 to analyse how policy instruments can lead to greening of the markets and stimulate more sustainable consumption patterns by individuals and households. She is the TNO project leader on BARENERGY, an EU project to identify the relevance and strengths of various barriers for energy behaviour changes among consumers and households. She was involved in the TNO programme New Initiative for Sustainable Innovations (NIDSI) on sustainable mobility.

Andrea Farsang is a PhD candidate and research affiliate at the Central European University (CEU), Department of Environmental Sciences and Policy, Hungary. Her research focuses on sustainable consumption and lifestyles, behaviour change and sustainability communication. After finishing her studies at St István University in Hungary and Hohenheim University in Germany, she worked as a researcher and project manager. At the CEU, Farsang has worked on several multidisciplinary research projects (FP6 and FP7) dealing with sustainability issues, energy efficiency and sustainable consumption. Her most recent work, the FP7 project Creating Innovative Sustainability Pathways (CRIPS), focuses on the identification of potential pathways to aid the EU towards the transition to a sustainable, low carbon Europe.

Shane Fudge is Lecturer in Energy Policy at the University of Exeter, UK. He is based in the Geography Department and teaches and supervises at undergraduate and postgraduate levels. From a background in sociology and politics, Fudge first became involved in energy and sustainability issues at the University of Surrey in 2006. Projects that he has been involved in since beginning work in the area of sustainability include: RESOLVE (research group on lifestyles, values and environment; BARENERGY (barriers and opportunities to changing consumer behavior at EU level); UNLOC (understanding local and community governance of energy); CRISP (creating innovative sustainability pathways); and REDUCE (reshaping energy demand of users by communication). He has also been involved in consultancy work, the latest one commissioned by The Association of Manufacturers of Domestic Appliances (AMDEA) into the relationship between technology and behavioural practices around household energy use. Fudge comes from a non-traditional academic background and is on the Board of Directors for the Surrey Lifelong Learning Partnership.

Isabelle Garabuau-Moussaoui has a PhD in social anthropology and has been working as a social sciences researcher at EDF (Electricité de France) R&D for nine years. She is specialised in socio-anthropology of consumption, daily habits and energy-related practices of households and of employees in companies. She is also involved in the UK CISE Project (Community Innovation for Sustainable Energy), funded by ECLEER and EPSRC. She is a trainer for adult classes in different Schools and is a member of the Network Sociology of Consumption and Uses at the French Association of Sociology (AFS). She is series editor at L'Harmattan. In April 2011 she published an article in English in the journal *Energy Efficiency*, 'Energy-related logics of action throughout the ages in France: historical milestones, stages of life and intergenerational transmissions'.

Cheryl Hicks is the team leader for sustainable lifestyles at the Collaborating Centre on Sustainable Consumption and Production (CSCP). At the management level Hicks directs the CSCP's projects on sustainable lifestyles. She has led the EU FP7 Social Platform project, SPREAD Sustainable Lifestyles 2050, which has developed a vision and roadmap for sustainable lifestyle models of the future, as project director. From 2004–10, Hicks worked with leading multinational corporations as the focal point for the World Business Council for Sustainable Development's (WBCSD) work on sustainable consumption and sustainable finance contributing to the WBCSD's 2008–09 scenarios project, Vision 2050. From 2001–04 she focused on the health agenda, and a process for dialogue related to corporate responsibility and impacts on health. From 1995–2001 she held various corporate positions in the areas of sales, marketing, e-business and human resources.

Angela High-Pippert is Associate Professor of Political Science and Director of the Women's Studies program at the University of St. Thomas in St. Paul, MN, US. Her research interests include citizen participation in community energy initiatives and women and politics, with publications in *Energy Policy*, *Carbon Management*, *Bulletin of Science, Technology and Society*, and *Women and Politics*. She is also co-editor of *Perspectives on Minnesota Government and Politics* (with Steven M. Hoffman and Kay Wolsborn). High-Pippert teaches courses in American politics, women and politics, and public policy, and received her university's Distinguished Educator Award in 2005 and 2012. She is past president of the Minnesota Political Science Association.

Steven M. Hoffman is Professor and Chair of the Department of Political Science at the University of St. Thomas, St. Paul, MN, US. He has published books, journal articles and research studies on community-based energy systems, environmental policy and, most recently, national identity in the newly independent states of East Central Europe. His latest work includes *Power Struggle: Hydro Development and First Nations in Manitoba and Quebec* (with Thibault Martin, 2008) and the last several editions of *Perspectives on Minnesota Government and Politics*. Hoffman has taught courses in environmental and energy policy, urban studies and comparative politics since his arrival at St Thomas in 1987, as well as study abroad and off-campus courses in New Zealand, East Central Europe and the Boundary Waters Canoe Area Wilderness. He has served as a director for a number of local environmental organizations, including Minnesotans for an Energy Efficient Economy, the Higher Education Consortium for Urban Affairs, Clean Water Action, Minnesota and Friends of the Boundary Waters Wilderness. He was recently awarded a University Scholars Grant to undertake research on the rapidly expanding and global exploitation of tar sands.

Michael Kuhndt is Head of the Collaborating Centre on Sustainable Consumption and Production (CSCP). At the management level, he presently directs projects in the fields of sustainable consumption and production and IPP, sustainability performance assessment and management, corporate responsibility and reporting, technology assessment, triple bottom line innovation, product stewardship and the design of strategies based on multi-stakeholder approaches at company, value chain and sectoral levels. Among other projects, he directed the development of the Pro Planet hotspot methodology for the German retailer REWE. Kuhndt has worked with or for a variety of organizations including the United Nations Development Programme (UNEP), the United Nations Industrial Development Organization (UNIDO), the World Bank, the International Labour Organization (ILO), the Global e-Sustainability Initiative, several European retailers and various consumer goods companies and business sector associations. He is a member of the Global Council on Sustainable Consumption at the World Economic Forum and co-chair of the Task Force on Sustainable Consumption and Green Development at the China Council for International Cooperation on Environment and Development.

Helma Luiten MSc, has been working at TNO, the Netherlands since 1998. Currently she combines her research and advisory work in the field of sustainable innovation with the function of Corporate Social Responsibility (CSR) officer. Luiten studied industrial design engineering at the Technical University of Delft. From environmentally friendly product development and innovation she developed towards sustainable system innovations and transitions. From a designers viewpoint she looks at sustainable innovations, that is, thinking ahead with an open mind and giving concrete form to ideas and consequences. She developed future visions for organizations and facilitated strategy development towards future visions using scenario development, roadmapping and learning trajectories. Luiten has experience with EU projects including the BarEnergy project (2008–10), EMUDE (project leader for TNO, 2004–06) and HiCS (2000–04). In her function as CSR officer she uses her knowledge to lead the implementation of sustainability within the TNO organization.

Ezio Manzini is Professor of Design at the Politecnico di Milano, Italy. Currently, he is Honorary Guest Professor at Tongji University, Shanghai, China, Jiangnan University, China Wuxi and at the COPPE-UFRJ, Rio de Janeiro, Brazil. In 2012 he was Distinguished Visiting Professor at Parsons, the New School for Design, New York, US. The most recent award granted to Manzini is the Sir Misha Black Medal in 2012. Throughout his professional life he has been at the Politecnico di Milano. Parallel to this, he collaborated with several international schools, from Domus Academy (in the 1990s) and to Hong Kong Polytechnic University (in 2000).

Scott Milne is Associate Research Fellow at the Surrey Energy Economics Centre, where he has contributed to scenario analysis of future energy systems and low carbon lifestyles as part of the EPSRC-funded Challenging Lock-in through Urban Energy Systems (CLUES) and ESRC-funded Research Group on Lifestyles, Values and Environment (RESOLVE). Milne was awarded his PhD in 2011 at the University of Surrey, exploring the carbon intensity of UK household consumption through to 2030.

Peter Newman is Professor of Sustainability at Curtin University in Perth, Australia. He is on the Board of Infrastructure Australia and is a Lead Author for Transport on the International Panel On Climate Change (IPCC). His books include *Green Urbanism in Asia* (2013), *Resilient Cities: Responding to Peak Oil and Climate Change (2009), Green Urbanism Down Under (2009)* and *Sustainability and Cities: Overcoming Automobile Dependence* (with Jeff Kenworthy), which was launched in the White House in 1999. From 2001–03 Newman directed the production of Western Australia's Sustainability Strategy in the Department of the Premier and Cabinet. In 2004 he was a Sustainability Commissioner in Sydney advising the government on planning and transport issues and in 2006 and 2007 he was a Fulbright Senior Scholar at the University of Virginia, Charlottesville, US. In late 2011 Newman was awarded the Sidney Luker medal by the Planning Institute of Australia (New South Wales) for his contribution to the science and practice of town planning in Australia.

Michael Peters is Lecturer in Energy Policy in the Built Environment in the School of Construction Management and Engineering at the University of Reading, UK, an appointment which began 1 October 2012. His research and teaching focus on energy governance and the policy and practice of reducing energy demand, with a special interest in community-level activities for addressing the complexities of climate change locally. Prior to joining Reading he worked at the University of Surrey, both as a Senior Research Fellow and as a Lecturer in the Centre for Environmental Strategy (CES). This included coordination of research in the ESRC Research Group on Lifestyles, Values and Environment (RESOLVE) on environmental education and community engagement in low carbon social change initiatives. Peters has extensive experience in the use of stakeholder research methods (including focus groups, deliberative workshops and interview techniques) and has presented his work on community engagement in climate change and sustainable energy policy widely at national and international conferences. He was lead editor for a recently published volume entitled *Low Carbon Communities: Imaginative Approaches to Combating Climate Change Locally* (Edward Elgar, September 2010).

Lucia Reisch is Professor for Intercultural Consumer Behaviour and Consumer Policy at the Department of Intercultural Communication and Management, Copenhagen Business School, Denmark. She is also an Honorary Research Professor at the German Institute for Economic Research (DIW) in Berlin and the Zeppelin University Friedrichshafen, Germany. An economist and social scientist by training, she holds a doctorate degree in economics. Her main research focus is on sustainable consumption, intercultural consumer behaviour, consumers and new technologies, consumer policy issues, behavioural economics and corporate sustainability. She is currently involved in several large-scale research projects on consumer behaviour and policy (EU FP6 and FP7) as well as in German and Swedish research projects on sustainable consumption and sustainability policy. She is editor in chief of the Journal of Consumer Policy and has more than 200 scientific publications to her name. As a policy consultant she serves on national and international advisory boards. Among various posts she holds is one as a member of the German Council for Sustainable Development, consulting the German government. After the Fukushima event in Japan she was invited, by German Chancellor Angela Merkel to be one of 12 members of the 'Ethics Commission' to advise the German government about Germany's energy future and its pathway regarding nuclear energy.

Eivind Stø received a Mag. art. in political science from the University of Oslo, Norway in 1972. He is presently the Director of Research at the

National Institute for Consumer Research (SIFO). Before joining SIFO in 1989 he was an assistant at the Norwegian Election Programme (1972–76) and a researcher at the Norwegian Fund for Market and Distribution Research (1976–89). He has initiated, participated in and coordinated several European projects. His research interests include consumer complaints, consumer policy, sustainable consumption, energy use and nano technology.

Pål Strandbakken received a Mag. art. in sociology from the University of Oslo, Norway in 1987 on a thesis on the Protestant ethic theory, and a PhD from the University of Tromsø in 2007 on a thesis on product durability and the environment. He has been working at the National Institute for Consumer Research (SIFO) since 1992. His research interests include product durability, energy use, ecological modernization and democratization of science.

Yolande Strengers is Vice Chancellor's Research Fellow at the Centre for Urban Research within the School of Global, Urban and Social Studies at RMIT University, Melbourne, Australia where she leads the Beyond Behaviour Change (BBC) Research Programme with Cecily Maller. Strengers' research is currently clustered around a series of applied projects that draw on theories of social practice to understand the dynamics of social and environmental change, and possibilities for intervening in the trajectories of practices. Energy demand management is a key topic of her research, particularly the impact of new smart technologies and associated strategies on household domestic practices. She holds a PhD in social science (RMIT University), a Masters in Social Science (RMIT University) and a Bachelor of Arts (Monash University). Prior to academia, she worked in the energy and sustainability sectors in communications and media management positions.

Martin van de Lindt joined TNO in 2004 as a senior consultant and researcher on system innovation, transition and transition management within the Strategy and Policy Group. He holds an MSc in human geography from Utrecht University, the Netherlands, with a specialization in regional and urban development, housing and planning, sociology and research methods and techniques, including market research. In his work he applies the concepts of transitions and transition management in different types of projects concerning sustainable energy, the built environment, mobility and consumer behaviour. Besides exploring the relationship between individual behaviour and system behaviour, he is concerned with the development of new methods and techniques based on system and transition thinking. Before joining TNO in 2004 van de Lindt already had

25 years of working experience, during which time he carried out a large number of research and consultancy projects for several Dutch ministries, provinces, municipalities, housing associations and construction companies. On behalf of the Dutch Ministry of Foreign Affairs he advised the cities of Prague and Brno in the Czech Republic, and Wrozlaw in Poland for several years on real estate maintenance/refurbish policy and on developing a policy for a sustainable built environment.

Walter Wehrmeyer is Reader in Environmental Business Management at the Centre for Environmental Strategy, University of Surrey, UK. His research interests include organizational approaches to innovation and sustainable development, participatory approaches to decision making and foresighting/backcasting as national strategies towards long-term change, including measurement systems to support benchmarking in these areas. Wehrmeyer was Director of the Entrepreneurship in Technology, IT and Business undergraduate programme, and is Adjunct Professor at the Graduate School of Business, Curtin University of Technology, Perth, Australia and a full member of the Institute of Environmental Management and Assessment. He is general editor of *Greener Management International* and is on the editorial board for the *Journal of Industrial Ecology*. He was awarded the University of Surrey's Vice Chancellor's Award for Teaching Excellence in 2011.

Foreword

Unless you are one of the people called 'climate change sceptics' – actually 'deniers', in that they reject rather than evaluate the obvious empirical evidence that human activities are changing the global climate – you recognize that human society needs to rethink the way we live and the economic activities that support our way of life. As an engineer, I have some understanding of how far technological developments can reduce the environmental impacts of consumption and I am sure that behavioural change is at least as important as technological change if we are to find the phantom called Sustainability. My engineering perspective also tells me that analysis, whether of global climate change or of consumer behaviour, is not enough: we need to go beyond understanding how individuals behave and what influences behaviour in different societies and contexts, to apply that understanding to promote behavioural change. The contribution of this book lies in this difficult area. It brings together analyses and case studies from different countries and disciplinary perspectives – psychological, sociological and economic – to provide an understanding of consumer behaviour and the barriers and inertia that constrain behavioural change, but goes further to inform policy and actions at all levels: local, national and supranational. It should therefore be useful not just to academic researchers but also anyone trying to promote changes in the behaviour of individuals and communities.

Policy concerns include being clear on objectives and also understanding what policy measures can and cannot achieve. Most readers of this book will be familiar with the 'tripartite' model of sustainability: the need to find a future that lies within the constraints imposed by the finite carrying and resource capacity of the planet, the need for technical and economic efficiency and the imperative of 'ensuring a better quality of life for everyone, now and for generations to come' (to quote from a UK government document). As an engineer, I sometimes feel obliged to point out that the 'laws' of economics are human constructs and therefore subject to policy intervention, whereas the laws of thermodynamics really are immutable and indulgences against thermodynamic laws are not available. The extent to which behaviour can be changed by policy interventions is a different matter again, not so rigid as thermodynamics nor yet so susceptible to direct intervention as the economic system. These simple observations underline why multidisciplinary approaches are essential: roads to sustainability do not lie within any single disciplinary field.

This 'flash of the blindingly obvious' was what prompted me to set up the Centre for Environmental Strategy (CES) at the University of Surrey, more than 20 years ago. It seemed to me (and unfortunately still looks that way) that much of the work published under the heading of Socio-Technical Studies was of limited merit or value because the 'socio' researchers had at best a tenuous grasp of technology and the technologists did not really appreciate the social sciences. The CES was intended to provide an environment where engineers, natural scientists and social scientists would work together directly, with a proper understanding not just of each other's terminology but each other's assumptions and perspectives. The measure of success in this kind of transdisciplinary work is whether you can address questions that cannot even be articulated within a single academic discipline. This book meets that criterion: it approaches the challenges of achieving sustainable lifestyles from different and multiple disciplinary perspectives. As an engineer who tries not to look at the world through a disciplinary tube, I value that. I hope you, the reader, will also find it valuable.

Roland Clift

Acknowledgements

The editors would like to thank all of the contributors and reviewers for their expertise, discernment and amiable cooperation. The publishers, Edward Elgar, are gratefully acknowledged for their kind support and assistance from the proposal development through to the finished product.

We would also like to pay tribute to the Centre for Environmental Strategy (CES) at the University of Surrey where three of the editors and several of the contributors to this volume have worked and with which many of the others have had close associations. Indeed, the central ideas for the volume were formulated by the lead editor during his time there. Now in its 21st year, the CES is an internationally acclaimed centre of excellence on sustainable development. It takes an interdisciplinary approach to the analysis of sustainable systems, integrating strong, engineering-based approaches with insights from the social sciences to develop action-oriented, policy-relevant responses to long-term environmental and social issues. Because the book has strong connections with the CES, we invited Roland Clift CBE, who founded the centre and is currently Executive Director of the International Society for Industrial Ecology, to contribute the Foreword.

Introduction

Shane Fudge, Michael Peters, Steven M. Hoffman and Walter Wehrmeyer

While a strengthened focus on the regulation of consumerism has been apparent at least since the 1987 Brundtland Report (Brundtland, 1987) the struggle to engage a uniform policy response around sustainability has been noticeable, not only at the level of the individual, but also at the systemic level of local, national, regional and international politics. The variable success of policies that have attempted to isolate and target agency as levers for change and transition suggests that behaviour-based policies must also be recognized as constituting part of the move towards wider, systemic change. As Fudge and Peters (2011) have argued, the scale and complexity of sustainability suggests that 'transition pathways' will only realistically come about as the result of a more negotiated shift that is played out through collaboration between governments, businesses, communities and individuals.

The result of this change in attitude has been an evolving focus on modifying individual and group behaviour as a more integral element in the design of policy strategies. In order to achieve meaningful reductions in energy-related carbon dioxide (CO_2) emissions, for instance, policy makers have begun to embrace a demand management approach, whereby lifestyle trends and patterns of individual and cultural consumption have been incorporated into more 'bottom-up' policy approaches. In many cases, this approach has been occasioned by the inability of conventional, mainly supply-side, energy strategies to achieve the sorts of long-term and enduring behavioural change required on the part of either households or firms. For instance, current estimates suggest that approximately 40 per cent of carbon emissions are embedded in household energy use and transport and mobility practices, both of which are often marginally affected by the sort of centralized policies and practices that have traditionally characterized energy planning. As Peters et al. (2012) have argued, such 'top-down' solutions will be effective, at best, only to the degree that they are able to engage with 'place-based' practices and behavioural norms, including those around food, energy, waste and other consumption habits and practices.

This book is premised on the belief that fresh ideas, new insights and interesting perspectives relevant to behaviour change within the context of the sustainability debate continue to be of paramount importance. Consider, for instance, the issue of climate change. Historically, emission-reducing policies have focused on regulating large and readily identifiable contributors to greenhouse gas levels such as power plants and other so-called point sources. While a number of countries have forced significant reductions, at least on a plant-by-plant basis, the limits of this approach are quickly being realized. As a result, effective future-oriented climate policy must be based on an all-together different approach, namely, one that focuses on the myriad ways in which individuals – as a category in their own right – put strains on the 'natural world' through long-practised consumption behaviours.

Changing behaviour is, of course, no simple task; the conflicts and divergent politics that exist around climate change again illustrate the difficulties in brokering national, much less international, agreement on how to design and implement a behaviour-based policy approach. Indeed, there often exists significant disagreement about whether behaviourial change is necessary or even desirable. China, for instance, has made the argument that its citizens have less of a responsibility to act on climate change than American citizens, where the average consumer emits many times the carbon on a per capita basis than her Chinese counterpart. In Sweden, on the other hand, large numbers of citizens themselves are influencing political action on environmental matters, while the UK government is only now beginning to realize that it will not reach stringent targets on greenhouse gas emissions without the willing engagement of its citizens.

Building upon a wide range of international contributions from academics, practitioners and experts in the area of behaviour change and sustainability, this book contributes to a theoretical understanding of both concepts. At the same time, the book provides numerous examples of past and current behaviour-based initiatives from around the world, many of which have been developed within diverse economic, cultural, social and political settings. In combining the theoretical with the practical, the book provides both academics and practitioners with a better understanding of what is and is not likely to resonate with individuals as they go about the uncertain business of choosing among energy options and behavioural alternatives.

ORGANIZATION OF THE BOOK

There is little doubt that the act of connecting with people, and the household/community configurations in which they find themselves, requires recognition of the fact that all individuals are inherently different. Indeed, if decision makers at any level, from the local to the international, are to make progress it is imperative that they acknowledge the full gamut of influences that shape, manipulate, encourage and deter consumer conduct and choice with regard to home energy and carbon management (Whitmarch, 2008). Part I of the book therefore provides a series of theoretical and methodological understandings of behaviour change, examining in particular the ways in which consumption-based approaches to sustainable development have generated an intensified focus on the role of citizens (individually and collectively) in policy making. In Chapter 1, Abrahamse and de Groot set out a thorough overview of the many contributions that psychology can make to the important issue of encouraging pro-environmental behaviour change. The chapter considers a range of factors that are related to behavioural choices, such as attitudes, habits, norms and values, highlighting their importance in providing insights into the process of behaviour change. It is argued that behaviour change initiatives and environmental policies need to be grounded strongly in psychological theories in order to enhance their effectiveness.

Some of the shortcomings of both psychological and economic theories provide a focal point for Chapter 2, which reframes the issue of peak electricity demand using theories of social practices. Contending that the 'problem' is one of transforming technologically-mediated social practices, Strengers reflects on how this body of theory repositions and refocuses the roles and practices of professions charged with the responsibility for affecting and managing energy demand. The chapter outlines three areas where demand managers could refocus their attention: (1) enabling co-management relationships with consumers; (2) working beyond their siloed roles with a broader range of human and non-human actors; and (3) promoting new practice 'needs' and expectations. The importance of identifying and establishing a new group of change agents is highlighted; agents who are actively but often unwittingly involved in reconfiguring the elements of problematic peaky practices.

It is widely recognized that the concept of rational choice dominates modern mainstream economic treatment of consumer behaviour. In Chapter 3, Milne charts the historical development of how this came to be, examining the methodological transition away from an early deductivist approach focused on assumptions towards a more instrumentalist approach focused on prediction, which resulted in the marginalization of concerns about the 'realism' of assumptions. The chapter draws on a recent econometric study that pried apart the relative contribution of income, price and non-economic factors in influencing changes in UK household expenditure over a 45-year period. The implications of this study are discussed in terms of the efficacy of economic interventions in influencing consumer behaviour, particularly in those categories of goods and services where non-economic factors are found to play the dominant role in explaining changes in expenditure.

A key area of consumer behaviour that has attracted increased attention for policy intervention in recent times – both in relation to health and environmental considerations – is that of food consumption. As Farsang and Reisch (Chapter 4) point out, considerable changes in food consumption, such as eating habits, dietary changes, availability and accessibility of food, have taken place over the last few decades. These are mainly due to an increase in productivity of the food sector, a greater diversity in product choices and a decrease in seasonal dependency due to global trade and storage and process technology. The food system is, however, a complex socio-ecological system surrounded by unpredictable events and uncertainties. The chapter explains how scenario planning is increasingly applied in both policy making and knowledge brokerage to deal with those uncertainties, complexities and long-term challenges as well as to influence developments proactively. The potential roles and applications of different types of scenarios such as visioning exercises, back-casting or quantitative models and their benefit for policy making are discussed. This is a field where ethical values feed into the consumption choices of consumers in a much more direct way than other areas, making this an interesting chapter to finish the first part of the book.

In Parts II and III, attention turns to the practical applications and political challenges that can surround attempts for promoting behaviour change at local, national and international levels. Part II focuses on different national interpretations at the European level and how behaviour change applications and interpretations relate to different cultural, political and economic influences.

Chapter 5 examines the new evidence or groundswells of promising sustainable lifestyle alternatives emerging across Europe. Hicks and Kuhndt describe a vision for more sustainable ways of living, review possible drivers of citizen action and propose a framework to analyse emerging trends towards more sustainable living. A number of promising practice examples are profiled according to the analytical framework. The authors argue that it will be important to continue to understand, test and evolve those concepts to ensure optimal positive impacts, and the consideration of possible unintended consequences or rebound effects.

Some of the key themes covered in Chapter 5 are then picked up by van de Lindt, Emmert and Luiten (Chapter 6) who report empirical findings from a three-year EU Framework Programme 7 study (BarEnergy) that explored the relevance and strength of barriers to energy changes amongst end consumers and households in six European countries. The project aimed to identify trends and influences in household energy use within and across the UK, France, Hungary, Switzerland, Norway and the Netherlands. The chapter highlights the potential influence of different stakeholder groups as effective intervention points within the countries themselves and also within an emerging pan-European environment/energy policy framework.

The focus for Manzini (Chapter 7) is on collective endeavour at the local level and the exciting possibilities that surround grassroots innovation, social change and enabling strategies. The chapter considers emerging collaborative (sustainable) behaviours and the enabling strategies capable of supporting them towards the achievement of mainstream status, as opposed to their current status as predominantly active minorities. The chapter therefore highlights the potential of niche initiatives to both engage with and transform behavioural norms from effective points of influence.

The important role of local-level activity in promoting more sustainable forms of social organization and environmental practices links closely with the principles and theory of sustainable consumption. In Chapter 8, Brugidou and Garabuau-Moussaoui describe how in France the issue of public policies relating to changes in sustainable consumption has recently been raised, providing a point of focus during the '*Grenelle de l'Environnement*', a consultancy and decision-making system bringing together, for the first time, environmental non-governmental organizations (NGOs), companies, trade unions, members of parliament and government officials. The question of how behavioural changes have been driven by energy-saving policies in France is examined, together with an exploration of the many social logics either favouring or restricting energy-saving practices, both within 'concerned' social groups and among the general public.

Reducing the environmental impact of households and consumers forms a key focal point for Strandbakken and Stø (Chapter 9) where the current Norwegian context is examined. The authors consider the issue of decoupling environmental impact from economic growth in Norway by critically evaluating the importance of recent trends in domestic energy consumption, meat consumption and household waste. These trends are discussed in the context of decoupling with possible rebound effects considered. The debate is framed in the sociology of consumption and it is stipulated that any signs of a reduced environmental impact from Norwegian households/consumers are potentially interesting because it could mean that it is possible to break the relationship between affluence and environmental impact.

Part III of the book explores in detail a range of initiatives and cases that have been developed in different countries on a more international scale around behavioural change and sustainable lifestyle agendas. In Chapter 10, for example, Ashton-Graham and Newman examine the strategy of 'sustainability coaching' as an effective approach to mobilizing large-scale behaviour change, described and discussed in relation to Australian households. This case study chapter presents the methods, results and findings from the Living Smart Households project – an initiative developed by the Department of Transport in Western Australia to build upon successful behavioural interventions in small group sustainability and large-scale transport and water demand management. The authors demonstrate that a process of self-framing allows individuals with different environmental, monetary and practical attitudes to adopt pro-environmental behaviours. It is also shown that commonly held values (such as the importance of water conservation to West Australians) can be utilized, through the coaching practice of engaging households in expressing their values (self-framing), to trigger behaviour change in unrelated areas such as energy, waste and travel choices.

In Chapter 11, Bhattacharyya considers the implications of macro-level changes on India's future energy needs. The chapter focuses on residential energy demand and uses the 'end-use' approach of demand analysis to evaluate possible effects. One of the key findings presented is that lifestyle changes are likely to have a considerable impact on future energy demand in India, but there is also the opportunity to follow a low-carbon pathway by adopting 'smart' technologies and creating efficient infrastructure. This, Bhattacharyya argues, requires a coordinated effort at various levels.

The difficulties and contradictions surrounding how the term 'community energy' can, or indeed should, be defined and used comes clearly into focus when considering a variety of such local-level initiatives and programmes in the USA. As Hoffman and High-Pippert (Chapter 12) discuss, decentralized or distributed forms of generation that are locally produced and/or consumed necessitate action at the local and community levels and the involvement of actors that can stimulate significant behavioural changes on the part of both households and local businesses. As the authors point out, however, in many cases 'community energy' initiatives are most appealing as rhetorical devices, useful for creating public support for otherwise controversial electrical projects. Drawing on a range of initiatives in the USA, the chapter addresses the question of how it might be possible to move beyond this situation when developing localized energy programmes.

In conclusion, the contributions to this book highlight both the promise and challenges of encouraging sustainable lifestyles as an intrinsic element of the changes required for society to adapt to the constraints of the planet. Taken together, the theoretical, practical and political insights provided set out a useful and up-to-date reference point in terms of the similarities, discrepancies, opportunities and barriers of different countries' attempts to embed behaviour change more explicitly into their sustainable development policy strategies. The international case studies provide robust evidence to show that it is indeed conceivable to translate the theory through to practice and that it really is possible to reach and influence 'ordinary' people and households, in spite of their hectic lifestyles, myriad commitments and other day-to-day priorities. This is not straightforward however, and the well-rehearsed problem of how to connect with those in society who do not consider that lifestyle changes for the sake of environmental sustainability should have anything to do with them personally continues to pose a substantial challenge for policy makers, practitioners and community groups alike. It also requires differentiation of the concept of 'the public' into different roles, such as 'citizens', 'consumers', 'voters', and so on to understand levers, barriers, drivers and the decision-making context within which pro-environmental behaviour evolves.

The fact that information-intensive initiatives alone are not capable of achieving the scale of lifestyle change desired by the types of initiative described in the chapters of this volume has almost reached the status of 'received wisdom', backed up by an extensive and growing evidence base of failed projects and sustainability programmes that only attract very weak levels of participation and framed as the 'value-action gap'. Internationally, the politics of development also means that changing individual behaviour is not yet a priority at policy level for many national governments. From the findings and observations presented in this book, it is clear that while individuals will hold a much more central role in policies to promote sustainability, a much more broad-reaching and holistic set of approaches will be required in order to optimize the effectiveness of behavioural and lifestyle change programmes as a key element in the decision maker's tool box. This necessitates, above all else, a commitment to bringing out and incorporating the quiescent ideas, concerns, hopes and zeal of all stakeholders involved. Importantly, this must include householders and community members themselves – the people individually and collectively who will ultimately make or break any sustainable lifestyles agenda.

REFERENCES

Bruntdland, G. (1987), *Our Common Future*, New York: United Nations, World Commission on Environment and Development.

Fudge, S. and M. Peters (2011), 'Behaviour change in the UK climate debate: an assessment of responsibility, agency and political dimensions', *Sustainability*, **3** (1), 291–302.

Peters, M., S. Fudge, S.M. Hoffman and A. High-Pippert (2012), 'Carbon management, local governance and community engagement', *Carbon Management*, **3** (4), 357–68.

Whitmarsh, L. (2008), 'Behavioural responses to climate change: asymmetry of intentions and impacts', *Journal of Environmental Psychology*, **29**, 13–23.

PART I

Theoretical and methodological understandings of behaviour change: psychological, sociological and economic perspectives

1. The psychology of behaviour change: an overview of theoretical and practical contributions

Wokje Abrahamse and Judith de Groot

INTRODUCTION

Current global trends indicate that our impact on the environment is considerable; illustrated particularly in the ways in which carbon dioxide emissions through the combustion of fossil fuels have steadily increased over the past decade (IPCC, 2007). It is now generally understood that behaviour changes need to become a more central aspect of the move towards a low carbon society. For example, a survey of US households indicates that greenhouse gas emissions from households could be reduced by up to 20 per cent through behaviour changes (Dietz et al., 2009). In the environmental policy arena, the idea of encouraging 'behaviour changes' and 'lifestyle changes' in order to reduce carbon footprints is gaining more and more attention (DEFRA, 2008).

Social and environmental psychologists have explored various ways to encourage people to adopt environmentally-friendly behaviours in order to alleviate the effects of human impacts on the environment (see Swim et al., 2009). Much of this research has been designed to try to uncover the underlying factors related to behavioural choices, such as values, identity and social norms. Psychologists feel that these factors are important to examine because they provide insight into both behaviour and how the process of behaviour change might be made to come about. Therefore, there is a wealth of psychological literature on the effectiveness of interventions aimed at encouraging environmentally-friendly behavioural choices, such as information provision, commitment making and feedback provision.

In this chapter, we provide a comprehensive overview of the contribution of psychological research to the area of behaviour change. We specifically focus on how insights from psychological research can contribute to enhancing and furthering our understanding of the effectiveness of behaviour change interventions. We conclude the chapter by highlighting avenues for future research.

FACTORS RELATED TO PRO-ENVIRONMENTAL BEHAVIOUR

Pro-environmental behaviours include those behaviours that impact the environment as little as possible, such as recycling, energy conservation and sustainable food consumption (Steg and Vlek, 2009). In the psychological literature on understanding and encouraging pro-environmental behaviour, two lines of research can be clearly distinguished. On the one hand, some studies focus on changing psychological factors related to behavioural choices, such as knowledge and attitudes, under the assumption that changes in attitude will result in behaviour change (Ajzen, 1991). Information campaigns, for example, are generally aimed at increasing people's knowledge or awareness on a certain issue, and it is assumed that increased knowledge or awareness will lead to changes in behaviour. Another line of research examines how pro-environmental behaviour can be encouraged by activating or resonating with certain psychological constructs, such as existing values, norms and identity (without necessarily changing them). Constructs such as values are assumed to be relatively stable over time, and by activating certain values (for example, environmental values), pro-environmental behaviour may be encouraged. In this section, therefore, we provide an overview of three important constructs that have been studied in the psychological literature, namely, values, identity and social norms. We then discuss how these constructs have been used as part of interventions to encourage pro-environmental behaviours.

Values

Various scholars have argued that unwillingness to change environmental behaviour is often rooted in values (Dunlap et al., 1983; Naess, 1989). Schwartz (1992, p. 21) defines a value as 'a desirable trans-situational goal varying in importance, which serves as a guiding principle in the life of a person or other social entity'. This definition suggests a number of important features in the relationship between values and behaviour. According to this model: (1) values include beliefs about the desirability or undesirability of certain end-states; (2) they are rather abstract constructs that transcend specific situations; and (3) they serve as guiding principles for the evaluation of people, events and behaviours. Another key feature of values is that they are prioritized in importance. This aspect implies that when competing values are activated in a specific situation, choices are based on the value that is considered most important.

Values have a number of important advantages in studying pro-environmental behaviour. The total number of values is relatively small compared to the countless behaviour-specific attitudes and norms. Consequently, they provide an efficient instrument for categorizing individuals and groups (Rokeach, 1973). Values can be seen as a general building block that can then be used to predict more specific attitudes and behaviours (Seligman and Katz, 1996). Further, values are considered to be relatively stable, guiding principles in people's lives. As such, values can, for example, be used to explain or predict attitudes towards the environment, attitudes towards new sustainable technologies or attitudes towards environmental policies. Values can be an important way in which to predict such attitudes because they provide a more stable and relatively enduring evaluation of these so-called 'attitude objects' (Stern et al., 1995). Finally, as argued above, there is evidence that values influence pro-environmental behaviours (Thøgersen and Ölander, 2002). Therefore, the study of values provides an excellent starting point for both understanding and changing behaviours. Through influencing or activating certain values, psychologists suggest that it is possible to influence a range of environmental behaviour-specific attitudes, norms, intentions and behaviours (for example, Verplanken and Holland, 2002).

Three types of values are particularly relevant for understanding pro-environmental attitudes, norms, intentions and actions. These are egoistic, altruistic and biospheric values (see, for example, de Groot and Steg, 2007a, 2008, 2010; Steg et al., 2005; Stern, 2000). For example, individuals who strongly endorse egoistic values will particularly consider the costs and benefits of environmental actions for them personally, and act pro-environmentally when the perceived personal benefits of a particular behaviour exceed the perceived personal costs. On the other hand, individuals who prioritize altruistic values will focus on the perceived costs and benefits of actions for society, and will act in a more environmentally-friendly way if the overall social benefits outweigh the social costs. Finally, persons who prioritize biospheric values will base their choice to act pro-environmentally on the perceived costs and benefits in relation to nature and the environment, and will act pro-environmentally when these actions are believed to reduce environmental problems.

All three value types may provide a distinct motivation to act pro-environmentally. However, studies have generally revealed that individuals who hold more strongly altruistic, and particularly biospheric, values have a tendency to engage in more positive pro-environmental attitudes, norms, intentions and behaviours than those who more strongly prioritize egoistic values (for example, de Groot and Steg, 2008, 2010; Honkanen and Verplanken, 2004; Karp, 1996; Steg et al., 2005; Steg et al., 2011). Psychologists suggest that this is probably because most pro-environmental behaviours require individuals to restrain egoistic tendencies (for example, de Groot and Steg, 2008; Nordlund and Garvill, 2002).

It has been shown that values are more influential on behaviour when they are activated in a specific situation (for example, Maio and Olson, 1998; Maio et al., 2001; Verplanken and Holland, 2002). This is significant for changing behaviour as activating values does not result in a value change as values are relatively stable and enduring over time (Schwartz, 1992). However, it is possible to make values more salient, which will affect the way people process relevant value-congruent information and subsequently act upon such information (de Groot and Steg, 2009). Scholars assume that when environmental values are linked to or are integrated into our self-concept, acting pro-environmentally can influence a moral rather than a selfish self-concept. Research suggests that instilling a sense of morality is more likely to promote pro-environmental actions as people generally like to see themselves as morally right (Aronson, 1969; see also section on 'identity').

Another way to promote value-congruent actions is by providing cognitive support for one's values, that is, by encouraging people to provide reasons for why they hold certain values (Maio and Olson, 1998). Studies have shown that without this kind of cognitive support, people have difficulty generating counterarguments against messages attacking an endorsed value, which may result in value-incongruent behaviour and even a change in value priority. Hence, activating values that are more likely to support pro-environmental behaviours may be an effective way to promote such behaviours.

Identity

A growing body of research indicates that identity may be related to engagement in pro-environmental behaviours (for example, Sparks and Shepherd, 1992; Whitmarsh and O'Neill, 2010). Identity is generally defined as an individual's adoption of particular groups (for example, 'I am female') and/or traits ('I am honest') as part of their self-concept (Tajfel and Turner, 1986). People may choose to engage in certain behaviours because the meanings people attach to these behaviours are consistent with their self-concept. If environmental issues are an important part of someone's self-concept for instance, engaging in pro-environmental behaviours may confirm his or her identity as someone who cares about the environment. For example, Sparks and Shepherd (1992) found that people who more strongly identified with being 'green' (environmentally friendly) were more likely to buy organic produce than those for which 'being green' was less important. Research indicates that environmental identity is positively related to engagement in pro-environmental behaviours (for example, Abrahamse et al., 2009; Stets and Biga, 2007; Whitmarsh and O'Neill, 2010).

Identity can be an important motivation for behaviour change as people generally strive for consistency between their self-concept and their behaviour. Cognitive dissonance can arise when individuals become aware of inconsistencies between their attitudes and their behaviour, or inconsistencies between different beliefs (Festinger, 1957). Significantly, Verplanken and Holland (2002) found that making environmental values salient that were central to one's self-concept (as opposed to peripheral) was effective in encouraging pro-environmental choices. In other words, by making environmental identities salient, environmental behaviour changes may be encouraged.

Social Norms

Social norms are 'rules and standards that are understood by members of a group, and that guide and/or constrain human behaviour without the force of laws' (Cialdini and Trost, 1998, p. 152). In simpler terms, social norms are thought to be beliefs about what are common and accepted behaviours for specific situations (Schultz and Tabanico, 2009). Social norms are sometimes confused with personal norms, which refer to an individual's belief about their moral obligation to engage in the behaviour (Schwartz, 1977). However, while social norms refer to what other people think or do, personal norms are rules or standards for one's own behaviour (Kallgren et al., 2000).

The perceived social norm is regarded as an important determinant for explaining pro-environmental attitudes, intentions and behaviours (de Groot and Steg, 2007b; Schade and Schlag, 2003) where a stronger social norm results in behaviour that is more in congruence with the perceived norm. For example, research found that people were more inclined to travel by public transport than the car when they believed other people approved of this behaviour as well (de Groot and Steg, 2007b). However, just like values, social norms assume particular significance in influencing attitudes, intentions and behaviours when they are made salient in a specific situation.

According to the focus theory of normative conduct (Cialdini et al., 1991), social norms can guide behaviour when these norms are made salient or focal. Two types of social norms are distinguished in the literature (Cialdini et al., 1991). Descriptive social norms refer to beliefs about what other people do in a specific situation, while injunctive social norms refer to beliefs about what other people approve or disapprove of doing. For example, studies have found that people were more likely to litter in littered environments (that is, a descriptive social norm in favour of littering was salient), and that this occurred to a greater extent when people saw somebody else littering (that is, injunctive social norm in favour of littering; Cialdini et al., 1990). Thus, one way to make norms salient is by providing specific cues in certain environments or situations, for example, through information provision.

Both types of norms may be used in messages that seek to resolve or highlight situations characterized by high levels of undesirable conduct (that is, acting in an environmentally harmful way is perceived as the norm). However, research indicates that the use of descriptive social norms may have unintended effects (Cialdini, 2003). For example, information campaigns sometimes contain information about the seriousness of the issue (for example, how many people litter), thereby highlighting a descriptive social norm in favour of the undesirable behaviour (for example, littering is the normal thing to do). This portrayal may actually increase the undesired behaviour (littering) because it is perceived as the (descriptive) norm. In these cases, it would be better to focus on the injunctive norm, which would convey approval of the desired behaviour (for example, to pick up litter).

Research has also shown that people often underestimate the influence that social norms may have on their own behaviour. For example, Nolan et al. (2008) showed that providing normative messages about the conservation behaviours of other households was most effective in increasing actual conservation behaviour. However, participants perceived the normative information, compared to other messages, such as saving money and being socially responsible, as least effective in changing their energy conservation behaviours. Thus, using knowledge about social norms in intervention strategies to encourage pro-environmental behaviours may not always be effective as the optimal conditions for change may not be present.

Interventions to Encourage Behaviour Change

The question of how to encourage pro-environmental behaviours has been a long-standing topic of interest in psychological research and the increased urgency of climate change continues to highlight its policy relevance. In this subsection, we briefly discuss three interventions that have been extensively studied in the literature. We highlight how insights from psychological research have contributed to enhancing the effectiveness of these interventions, with a particular focus on how values, identity and social norms may be used in intervention strategies to encourage pro-environmental behaviours. More extensive overviews of the intervention literature can be found elsewhere (see, for example, Abrahamse and Matthies, 2012; Abrahamse et al., 2005; Dwyer et al., 1993; Schultz et al., 1995).

Information provision

Information provision is perhaps the most widely used intervention to encourage the adoption of pro-environmental behaviours. Information provision is rooted in the so-called knowledge-deficit model (Schultz, 2002), and works on the assumption that people do not act pro-environmentally because they have a lack of knowledge about environmental problems, and/or do not know in detail what they themselves can do to address these problems. However, while it has been shown that information provision generally results in increased knowledge and awareness, in many cases it does not necessarily translate into actual behaviour changes (Abrahamse et al., 2005).

Cialdini (2003) has argued that the effectiveness of information campaigns may be enhanced by incorporating insights from theories about social norms. As highlighted in the previous subsection, behaviour change may be encouraged or may be more likely when social norms are activated or when they are made salient in a particular situation. An increasing body of literature has focused on the extent to which the salience of social norms can spur pro-environmental behaviour. For instance, an often cited study by Goldstein et al. (2008), which focused on the effect of normative messages on encouraging towel reuse in hotels, found that towels were reused more frequently when hotel guests were provided with descriptive norm information regarding how many other guests were reusing towels, compared to the standard environmental message often used in hotels. Their research also suggested that towel reuse was higher when the message referred to a specific reference group (that is, hotel guests who had stayed in the same room) as compared to a general reference group (that is, hotel guests).

Research on values can also be helpful for designing information campaigns to promote pro-environmental behaviours. Research indicates that the activation of pro-environmental values can encourage people to make pro-environmental choices (Verplanken and Holland, 2002). For example, a campaign could focus on the specific biospheric consequences of reducing global warming or improving the local environment when trying to reduce commuting by car. By activating biospheric values in people, a shift towards more pro-environmental behaviours may be encouraged. Another way to utilize research on values is by integrating values into the provision of tailored information. Tailored information is designed to reach a specific person or group(s) of people on the basis of characteristics unique to those individuals (Kreuter et al., 1999). It has been found that the provision of tailored information is generally more effective at encouraging pro-environmental behaviours than the provision of general information (for example, Abrahamse et al., 2005; Abrahamse et al., 2007). As argued earlier, individuals differ in their value priorities and can therefore be divided into relatively homogeneous groups that can be targeted by different messages (for example, Kamakura and Mazzon, 1991).

Commitments

In a commitment intervention, individuals or groups are asked to sign a pledge (commitment) to change their behaviour. This commitment can be a pledge to oneself, but it can also be made public, for instance, by means of an announcement in the local newspaper. A recent meta-analysis indicated that commitment making is generally effective at encouraging pro-environmental behaviours (Lokhorst et al., 2013). There is some indication that public commitments may work differently in different social groups. Wang and Katzev (1990) found that residents of a retirement home who had signed a public commitment (at a group meeting) increased recycling both during the intervention as well as in a follow-up period. In a sequel to this study, students were either asked to sign a public commitment or an individual commitment. The individual commitment group recycled significantly more than the public commitment group.

There is still relatively little known of the underlying mechanisms of commitments and how commitment making helps encourage behaviour change (Lokhorst et al., in press). One possible explanation that has been put forward emphasizes the alignment between people's values, identity and behaviours. According to Cialdini (2001), people generally have a need for consistency, and when people make a commitment and subsequently stick to this commitment by changing their behaviour, they may start aligning their values or their self-concept with the changed behaviour. So, someone who has made a commitment to start recycling, and who then starts recycling, may start to see recycling as an important part of who they are or as an important part of their value system (see also Seligman and Katz, 1996). A study on water conservation examined the role of evoking cognitive dissonance and commitment making (Dickerson et al., 1992). Students were assigned to one of three conditions. In group 1, students were made aware of their own water wasting behaviour when showering. In group 2, students were asked to publicly commit to conserving water when showering. Group 3 received a combination of these two treatments, which was assumed to evoke cognitive dissonance between their awareness of wasting water and their commitment to conserving water. Those in the cognitive dissonance group had the shortest showering time.

Feedback provision

Feedback is an intervention that is often applied to encourage the adoption of pro-environmental behaviours, such as energy conservation (Abrahamse et al., 2007; Darby, 2006), recycling (Schultz, 1998) and water conservation (Aitken et al., 1994). The provision of feedback consists of giving people information about their performance, for instance, in relation to energy savings or amount of recycled materials. According to feedback intervention theory (Kluger and DeNisi, 1996), feedback may encourage behaviour change because it provides insight into the links between certain outcomes (for example, saving energy) and the behaviour changes necessary to reach those outcomes (for example, switching off lights).

One way to enhance the efficacy of feedback provision is to increase the salience of social norms (Schultz et al., 2007). For example, in a study on energy use, households received feedback about the average energy use of other households in their community (activating a descriptive social norm). In addition, some households also received feedback that indicated approval or disapproval of their energy usage as compared to the average by means of a happy or sad face (activating an injunctive norm). It appeared that the effectiveness of normative feedback depended on whether households had high or low energy usage. Households with above-average consumption reduced energy use as a result of descriptive norm feedback and in relation to a combination of descriptive and injunctive norm feedback. Below-average consumers, on the other hand, increased energy use as a result of descriptive norm feedback, while their energy usage did not change as a result of the combination of descriptive and injunctive norm feedback.

In another example, the principle of cognitive dissonance has been employed as part of feedback interventions to encourage behaviour change (for example, Aitken et al., 1994; Kantola et al., 1984). It was found that by making a discrepancy salient between people's beliefs and their behaviour, behaviour change may be encouraged. A study by Kantola and colleagues (1984) also showed that by giving feedback that evoked dissonance between people's reported attitudes (that is, towards energy conservation) and their actual behaviour (that is, high energy usage), households significantly reduced their energy use. As highlighted earlier in the chapter, the principle of cognitive dissonance has also been successfully used as part of other interventions, notably commitment making (Dickerson et al., 1992).

CONCLUSION

Psychology has important contributions to make in helping to both understand and alleviate environmental problems (Swim et al., 2009). In this chapter, we have provided a snapshot of psychological research that has examined the role of values, identity and social norms in relation to pro-environmental behaviour. We focused in particular on the effectiveness of activating values, identity and social norms in intervention research as a way to encourage pro-environmental behaviours.

When the aim is to encourage behaviour change amongst individuals, it is important to understand the factors related to behavioural choices. In general, a strong connection is needed between theory-driven research and applied intervention research. For instance, theoretical insights in relation to how social norms are related to behaviour can be used as part of interventions to encourage behaviour change (see Cialdini, 2003). In particular, making values, social norms and identity salient can be a viable route to encourage behaviour change. As we have highlighted, the activation of social norms encourages the adoption of various pro-environmental behaviours, such as energy conservation (Schultz et al., 2007) and towel reuse (Goldstein et al., 2008). Similarly, the activation of values and identity may encourage more widespread pro-environmental choices (Verplanken and Holland, 2002).

A number of important questions remain however. Firstly, there is still relatively little known about the underlying processes of behaviour change (Abrahamse et al., 2005; Lokhorst et al., in press). Future studies that examine the effect of activating values, identity and social norms could include measures of psychological variables to gain insight into the process of behaviour change. For example, studies that use social norm information could include a measure of social norms to assess whether the information was indeed effective at influencing social norms. A second important issue relates to the question of how a range of pro-environmental behaviours that people undertake in their daily lives can be encouraged. It has been suggested, for instance, that when people start engaging in one pro-environmental behaviour (for example, recycling), they may be inclined to start doing other things as well (for example, composting); referred to as the 'spill-over effect' (Thøgersen and Ölander, 2003). A possible explanation for this is that people may start to see themselves as those who care about the environment (they develop a 'green' identity or prioritize biospheric values), which in turn guides other behavioural choices. Further research is therefore needed to examine these suggested links between people's values, identity and engagement in (a range of) pro-environmental behaviours. Lastly, future research needs to address to what extent does the activation of factors such as values, identity and social norms lead to long-term behaviour change. What happens, for instance, when the intervention is discontinued? The evidence so far appears to be mixed. While there is some indication that commitment making has long term effects (Lokhorst et al., 2013), very few studies include follow-up measurements, and relatively little is known about long-term effects. Clearly, more research is needed that includes longer-term measures of behaviour change, which may help inform future environmental policies and other initiatives aimed at encouraging more widespread pro-environmental lifestyles.

REFERENCES

Abrahamse, W. and E. Matthies (2012), 'Informational strategies to promote pro-environmental behaviour: changing knowledge, awareness, and attitudes', in L. Steg, A.E. van den Berg and J.I.M. de Groot (eds), *Environmental Psychology: An Introduction*, Oxford: Wiley-Blackwell, pp. 223–32.

Abrahamse, W., B. Gatersleben and D. Uzzell (2009), 'Encouraging sustainable food consumption. The role of (threatened) identity', RESOLVE working paper series, 04–09, available at http://resolve.sustainablelifestyles.ac.uk/publications (accessed 24 June 2013).

Abrahamse, W., L. Steg, C. Vlek and T. Rothengatter (2005), 'A review of intervention studies aimed at household energy conservation', *Journal of Environmental Psychology*, **25**, 273–91.

Abrahamse, W., L. Steg, C. Vlek and T. Rothengatter (2007), 'The effect of tailored information, goal setting and feedback on household energy use, energy-related behaviors and behavioral determinants', *Journal of Environmental Psychology*, **27**, 265–76.

Aitken, C.K., T.A. McMahon, A.J. Wearing and B.L. Finlayson (1994), 'Residential water use: predicting and reducing consumption', *Journal of Applied Social Psychology*, **24**, 136–58.

Ajzen, I. (1991), 'The theory of planned behavior', *Organizational Behavior and Human Decision Processes*, **50**, 179–211.

Aronson, E. (1969), 'A theory of cognitive dissonance: a current perspective', in L. Berkowitz (ed.), *Advances in Experimental Social Psychology*, 4th edn, New York: Academic Press, pp. 1–34.

Cialdini, R.B. (2001), *Influence: Science and Practice*, Boston, MA: Allyn & Bacon.

Cialdini, R.B. (2003), 'Crafting normative messages to protect the environment', *Current Directions in Psychological Science*, **12**, 105–9.

Cialdini, R.B. and M.R. Trost (1998), 'Social influence: social norms, conformity, and compliance', in D. Gilber, S. Fiske and G. Lindzey (eds), *Handbook of Social Psychology*, 4th edn, Vol. 2, Boston, MA: McGraw-Hill, pp. 151–92.

Cialdini, R.B., C.A. Kallgren and R.R. Reno (1991), 'A focus theory of normative conduct: a theoretical refinement and re-evaluation of the role of norms in human behavior', in M.P. Zanna (ed.), *Advances in Experimental Social Psychology*, Vol. 24, San Diego, CA: Academic Press pp. 201–34.

Cialdini, R.B., R.R. Reno and C.A. Kallgren (1990), 'A focus theory of normative conduct – recycling the concept of norms to reduce littering in public places', *Journal of Personality and Social Psychology*, **58**, 1015–26.

Darby, S. (2006), *The Effectiveness of Feedback on Energy Consumption. A Review for DEFRA of the Literature on Metering, Billing and Direct Displays*, Oxford: Environmental Change Institute, University of Oxford.

De Groot, J.I.M. and L. Steg (2007a), 'Value orientations and environmental beliefs in five countries: validity of an instrument to measure egoistic, altruistic and biospheric value orientations', *Journal of Cross-Cultural Psychology*, **38**, 318–32.

De Groot, J.I.M. and L. Steg (2007b), 'General beliefs and the theory of planned behavior: the role of environmental concerns in the TPB', *Journal of Applied Social Psychology*, **37**, 1817–36.

De Groot, J.I.M. and L. Steg (2008), 'Value orientations to explain beliefs related to environmental significant behavior: how to measure egoistic, altruistic, and biospheric value orientations', *Environment and Behavior*, **40**, 330–54.

De Groot, J.I.M. and L. Steg (2009), 'Mean or green? Values, morality and environmental significant behavior', *Conservation Letters*, **2**, 61–6.

De Groot, J.I.M. and L. Steg (2010), 'Relationships between value orientations, self-determined motivational types and pro-environmental behavioural intentions', *Journal of Environmental Psychology*, **30**, 368–78.

DEFRA (Department for Environment, Food and Rural Affairs) (2008), *A Framework for Pro-environmental Behaviours*, London: Department for Environment, Food and Rural Affairs, available at http://www.defra.gov.uk/publications/files/pb13574-behaviours-report-080110.pdf (accessed 24 June 2013).

Dickerson, C.A., R. Thibodeau, E. Aronson and D. Miller (1992), 'Using cognitive dissonance to encourage water conservation', *Journal of Applied Social Psychology*, **22**, 841–54.

Dietz, T., G.T. Gardner, J. Gilligan, P.C. Stern and M.P. Vandenbergh (2009), 'Household actions can provide a behavioral wedge to rapidly reduce U.S. carbon emissions', *Proceedings of the National Academy of Sciences of the USA*, **106** (44), 18452–6.

Dunlap, R.E., J.K. Grieneeks and M. Rokeach (1983), 'Human values and pro-environmental behaviour', in W.D. Conn (ed.), *Energy and Material Resources: Attitudes, Values, and Public Policy*, Boulder, CO: Westview Press, pp. 145–68.

Dwyer, W.O., F.C. Leeming, M.K. Cobern, B.E. Porter and J.M. Jackson (1993), 'Critical review of behavioral interventions to preserve the environment. Research since 1980', *Environment and Behavior*, **25**, 275–321.

Festinger, L. (1957), *A Theory of Cognitive Dissonance*, Stanford, CA: Stanford University Press.

Goldstein, N.J., R.B. Cialdini and V. Griskevicius (2008), 'A room with a viewpoint: using social norms to motivate environmental conservation in hotels', *Journal of Consumer Research*, **35**, 472–82.

Honkanen, P. and B. Verplanken (2004), 'Understanding attitudes towards genetically modified food: the role of values and attitude strength', *Journal of Consumer Policy*, **27**, 401–20.

IPCC (Intergovernmental Panel on Climate Change) (2007), 'Summary for policymakers', *in Climate Change 2007: The Physical Science Basis*, contribution of working group I to the fourth assessment report of the Intergovernmental Panel on Climate Change, available at http://www.ipcc.ch/pdf/assessment-report/ar4/wg1/ar4-wg1-spm.pdf (accessed 24 June 2013)

Kallgren, C.A., R.R. Reno and R.B. Cialdini (2000), 'A focus theory of normative conduct: when norms do and do not affect behaviour', *Personality and Social Psychology Bulletin*, **26**, 1002–12.

Kamakura, W.A. and J.A. Mazzon (1991), 'Value segmentation: a model for the measurement of values and value systems', *Journal of Consumer Research*, **18**, 208–18.

Kantola, S.J., G.L. Syme and N.A. Campbell (1984), 'Cognitive dissonance and energy conservation', *Journal of Applied Psychology*, **69**, 416–21.

Karp, D.G. (1996), 'Values and their effect on pro-environmental behavior', *Environment and Behavior*, **28**, 111–33.

Kluger, A.N. and A. DeNisi (1996), 'The effects of feedback interventions on performance: a historical review, a meta-analysis, and a preliminary feedback intervention theory', *Psychological Bulletin*, **119**, 254–84.

Kreuter, M.W., D. Farrell, L. Olevitch and L. Brennan (1999), *Tailored Health Messages: Customizing Communication with Computer Technology*, Mahwah, NJ: Lawrence Erlbaum.

Lokhorst, A.M., C. Werner, H. Staats, E. van Dijk and J.L. Gale (2013), 'Commitment and behavior change: a meta-analysis and critical review of commitment-making strategies in environmental research', *Environment and Behaviour*, **45**, 3–34.

Maio, G.R. and J.M. Olson (1998), 'Values as truisms: evidence and implications', *Journal of Personality and Social Psychology*, **74**, 294–311.

Maio, G.R., J.M. Olson, L. Allen and M. Bernard (2001), 'Addressing discrepancies between values and behavior: the motivating effect of reasons', *Journal of Experimental Social Psychology*, **37**, 104–17.

Naess, A. (1989), *Ecology, Community, and Lifestyle*, Cambridge: Cambridge University Press.

Nolan, J., P.W. Schultz, R.B. Cialdini, V. Griskevicius and N. Goldstein (2008), 'Normative social influence is underdetected', *Personality and Social Psychology Bulletin*, **34**, 913–23.

Nordlund, A.M. and J. Garvill (2002), 'Value structures behind pro-environmental behavior', *Environment and Behaviour*, **34**, 740–56.

Rokeach, M. (1973), *The Nature of Human Values*, New York: The Free Press.

Schade, J. and B. Schlag (2003), 'Acceptability of urban transport pricing strategies', *Transportation Research Part F: Traffic Psychology and Behaviour*, **6**, 45–61.

Schultz, P.W. (1998), 'Changing behavior with normative feedback interventions: a field experiment on curbside recycling', *Basic and Applied Psychology*, **21**, 25–36.

Schultz, P.W., S. Oskamp and T. Mainieri (1995), 'Who recycles and when? A review of personal and situational factors', *Journal of Environmental Psychology*, **15**, 105–21.

Schultz, P.W. and J.J. Tabanico (2009), 'Criminal beware: a social norms perspective on posting public warning signs', *Criminology*, **47**, 1201–22.

Schultz, P.W. (2002), 'Knowledge, education, and household recycling: examining the knowledge-deficit model of behavior change', in T. Dietz and P. Stern (eds), *New Tools for Environmental Protection*, Washington, DC: National Academy of Sciences, pp. 67–82.

Schultz, P.W., J.M. Nolan, R.B. Cialdini, N.J. Goldstein and V. Griskevicius (2007), 'The constructive, destructive, and reconstructive power of social norms', *Psychological Science*, **18**, 429–34.

Schwartz, S.H. (1977), 'Normative influences on altruism', *Advances in Experimental Social Psychology*, **10**, 221–79.

Schwartz, S.H. (1992), 'Universals in the content and structures of values: theoretical advances and empirical tests in 20 countries', in M. Zanna (ed.), *Advances in Experimental Psychology*, Vol. 25, Orlando, FL: Academic Press, pp. 1–65.

Seligman, C. and A.N. Katz (1996), 'The dynamics of value systems', in C. Seligman, J.M. Olson and M.P. Zanna (eds), *The Psychology of Values: The Ontario Symposium*, Vol. 8, Hillsdale, NJ: Lawrence Erlbaum Associates, pp. 53–75.

Sparks, P. and R. Shepherd (1992), 'Self-identity and the theory of planned behaviour: assessing the role of identification with green consumerism', *Social Psychology Quarterly*, **55**, 388–99.

Steg, L. and C. Vlek (2009), 'Encouraging pro-environmental behaviour: an integrative review and research agenda', *Journal of Environmental Psychology*, **29**, 309–17.

Steg, L., J.I.M. de Groot, L. Dreijerink, W. Abrahamse and F. Siero (2011), 'General antecedents of personal norms, policy acceptability, and intentions. The role of values, worldviews, and environmental concern', *Society and Natural Resources*, **24**, 349–67.

Steg, L., L. Dreijerink and W. Abrahamse (2005), 'Factors influencing the acceptability of energy policies: testing VBN theory', *Journal of Environmental Psychology*, **25**, 415–25.

Stern, P.C. (2000), 'Toward a coherent theory of environmentally significant behavior', *Journal of Social Issues*, **56** (3), 407–24.

Stern, P.C., T. Dietz, L. Kalof and G.A. Guagnano (1995), 'Values, beliefs, and pro-environmental action: attitude formation toward emergent attitude objects', *Journal of Applied Social Psychology*, **25**, 1611–36.

Stets, J.E. and C.F. Biga (2007), 'Bringing identity theory into environmental sociology', *Sociological Theory*, **21**, 398–423.

Swim, J., S. Clayton, T. Doherty et al. (2009), *Psychology and Global Climate Change: Addressing a Multi-faceted Phenomenon and Set of Challenges*, report by the American Psychological Association's Task Force on the Interface between Psychology and Global Climate Change, American Psychological Association, available at http://www.apa.org/science/about/publications/climate-change-booklet.pdf (accessed 12 December 2011).

Tajfel, H. and J.C. Turner (1986), 'The social identity theory of intergroup behaviour', in S. Worchel and W.G. Austin (eds), *Psychology of Intergroup Relations*, Chicago, IL: Nelson-Hall, pp. 7–24.

Thøgersen, J. and F. Ölander (2002), 'Human values and the emergence of a sustainable consumption pattern: a panel study', *Journal of Economic Psychology*, **23**, 605–30.

Thøgersen, J. and F. Ölander (2003), 'Spill-over of environment-friendly consumer behaviour', *Journal of Environmental Psychology*, **23**, 225–36.
Verplanken, B. and R.W. Holland (2002), 'Motivated decision making: effects of activation and self-centrality of values on choices and behaviour', *Journal of Personality and Social Psychology*, **82**, 434–47.
Wang, T.H. and R.D. Katzev (1990), 'Group commitment and resource conservation: two field experiments on promoting recycling', *Journal of Applied Social Psychology*, **20**, 265–75.
Whitmarsh, L. and S. O'Neill (2010), 'Green identity, green living? The role of pro-environmental self-identity in determining consistency across diverse pro-environmental behaviours', *Journal of Environmental Psychology*, **30**, 305–14.

2. Peak electricity demand and social practice theories: reframing the role of change agents in the energy sector[1]

Yolande Strengers

INTRODUCTION

Peak electricity demand is a pressing international energy policy concern, causing widespread blackouts and increasing the cost of electricity for all consumers. In Australia alone, billions of dollars in investment are being used to upgrade electricity distribution and transmission infrastructure, and build generation plants to provide power during periods of peak demand. Despite these efforts (and in some cases because of them), there are growing concerns about the frequency of blackouts, particularly on hot summer days when residential air-conditioning demand adds disproportionately to peak loads (Wilkenfeld, 2004). Consequently, a range of demand management strategies have emerged, such as time-of-use pricing and consumption feedback, to educate and incentivize consumers to shift or reduce peak demand.

The primary purpose of this conceptual chapter is not to contribute towards existing debates about where demand management programmes and peak electricity investment would be best targeted, but rather to reframe the issue entirely. The focus is on how the 'problem' of peak electricity demand and the demand management 'solutions' it generates emerge from a particular construction of reality that places humans and their minds at the forefront of social order. This humanist perspective continues to dominate into the twenty-first century (Schatzki, 2001), and is the foundation for the production of knowledge, policy and programmes intended to achieve social and environmental transformation in an era of climate change and resource uncertainty (Shove, 2010). In the context of peak electricity demand, this construction is most evident in policies and programmes that attempt to 'shift' and 'shed' consumer demand.

The chapter departs from this dominant understanding of social order and change, instead drawing on social practice theories as 'a distinct social ontology', whereby 'the social is a field of embodied, materially interwoven practices' (Schatzki, 2001, p. 3). Social practice theories depart from accounts that privilege social totality (social norms), institutions or systems (structure), cultural symbols and meanings (symbolism) or attitudes, behaviours and choices. They also overcome common dualisms that manifest themselves in the energy and resource sectors, such as supply and demand, consumption and production or behaviour and technology. In this chapter, I demonstrate how they can reframe the issue of peak electricity demand as one of changing and shifting technologically-mediated social practices, resulting in different foci for demand managers assigned the role of affecting change.

This is not primarily a debate or discussion about theory, but about the ways in which theory 'works on' policy and manifests itself in energy strategy. While theories can only ever be abstractions and constructions of reality, they can and do have quite profound effects on it. As Shove (2011, p. 264, emphasis in original) argues, the value of alternative theories of social change is to 'generate *different* definitions of the problem', not to provide a more 'holistic' perspective or to solve existing policy (and resource management) problems. Alternative perspectives are particularly necessary for the energy sector, where a unified body of theory, research and practice has served to construct and reinforce clear knowledge, processes and policies for the task of managing demand.

In addition to recasting the peak demand problem, this chapter aims to identify what this reframing might mean for the professions charged with the responsibility and agency for steering demand, and how it might reorient the practice of doing and being a 'change agent' in the energy sector. Traditional change agents are most clearly exemplified in their roles as demand managers, where their primary task is to deploy a range of instruments such as pricing incentives and disincentives, educational and informational strategies (for example, consumption feedback) and technological solutions intended to shift or shed energy demand. For the purposes of this chapter I use the term 'demand manager' broadly to refer to a range of professions, such as load managers, consumer and customer relations teams, smart metering programme managers, behaviour change practitioners, energy-efficiency advisors and network pricing managers. While not all of these professions might identify themselves as demand managers, they are variously involved in attempting to steer, redirect or intervene in consumer demand through a range of programmes, incentives and technological intermediaries that they deliver and/or promote.

I begin by examining the self-reinforcing demand management paradigm and how it is employed to understand and frame energy problems, before presenting social practice theories as an alternative perspective in the third section. In the fourth section, I discuss the ways in which social practice theories potentially reposition the problem of peak electricity demand and the role of demand managers, before identifying three different foci for these professions in the fifth section. The sixth section attends to the different ways in which agency is assigned and agents identified between these different understandings of people and their demand, and what this means for who or what can be considered a change agent. I argue for the establishment and identification of a new breed of change agents who are actively but often unwittingly involved in reconfiguring the elements of problematic 'peaky practices'. However, I warn that assigning certain professions (but not others) with the agency to affect social change may be misleading and unhelpful in achieving it.

THE SUPPLY–DEMAND DIVIDE: A SELF-REINFORCING PARADIGM

In the energy sector, the dominant paradigm is one where supply is split from demand, with technological efficiency on one side and behavioural improvements on the other. Consumers are framed as rational, self-interested and autonomous agents, whereas technology is viewed as 'an impartial, instrumental tool in a "win-win" scenario that couples economic growth with environmental improvement' (Hobson, 2006, p. 319). This divided view encourages a two-tiered approach to energy management problems that prioritize separate supply-side and demand-side solutions.

In the case of peak electricity demand, focusing on the supply side by upgrading electricity infrastructure and generation capacity is often viewed as economically inefficient investment. In Australia, for example, this peak capacity is only required for 1–2 per cent of the year (ETSA, 2007) and causes widespread blackouts on hot summer days, leaving householders vulnerable to the effects of heat (Maller and Strengers, 2011; Strengers and Maller, 2011). While a range of household appliances are implicated in peaky practices, such as televisions, heaters, home computers, refrigerators, pool pumps, washing machines and dishwashers (Pears, 2004), the air-conditioner has attracted particular attention as a 'culprit' appliance, primarily as a result of its rapid diffusion and increasing penetration (DEWHA, 2008). Similar scenarios are playing out internationally, focusing demand managers' attention on air-conditioning (and other peak) load during peak periods (Herter, 2007; Strengers, 2010a).

Popular demand-side solutions include variable pricing regimes, consumption feedback and 'direct load control' (the remote control of appliances with a high load during peak times). These strategies are often facilitated through government mandates for smart metering (Darby, 2010). Additionally, governments, non-governmental organizations (NEOs) and energy utilities employ a range of behavioural strategies to curb demand, such as informative websites and books about how to save energy, and educational programmes and campaigns designed to assist people in making more resource-efficient decisions and investments about their consumption.

To undertake these tasks, demand managers draw on a unified collection of human-centred, psychological and economic theories, which Elizabeth Shove (2010) has termed the 'Attitudes, Behaviour, Choice' (ABC) model. While Shove originally posited and critiqued the ABC model as the foundation for strategies designed to encourage energy conservation and reduce greenhouse gas emissions, it is equally applicable to the other key objective of demand management, which is to shift demand to non-peak times of the day. When load shifting is the primary aim, the focus of the ABC model is expanded to include the transfer of demand management skills to energy consumers. Householders are expected to transform into micro resource managers (Strengers, 2011b), and are represented as 'Mini-Me' versions of their utility providers who must make similar resource management decisions at the household level (Sofoulis, 2011). The aim is to encourage consumers to make autonomous and cost-reflective decisions about the scheduling of their consumption (in accordance with their attitudes, behaviours and choices) through incentives and disincentives such as variable pricing tariffs. To highlight this additional emphasis of demand management, I graft a 'D' for 'Demand' onto Shove's ABC model for the purposes of this chapter.

This ABCD model pervades energy policy and management, and is reinforced by the plethora of consumer- and demand-oriented opportunities open to researchers, industry and community groups to respond to discussion documents and granting schemes. For example, an Australian Energy Market Commission review with the telling title 'Power of choice – giving consumers options in the way they use electricity' (AEMC, 2011), asked for responses to a series of predefined questions aimed at identifying consumer 'drivers', 'choices' and information needs. Here a series of epistemological and ontological assumptions about how humans understand their world, act within it, acquire new knowledge and instigate change are adopted from the outset. The emphasis is on changing and responding to what is going on in the minds of individual consumers.

Similarly, a recent surge of international reports on the 'consumer domain' (CEA, 2011), 'consumer impacts' (NERA, 2008), 'maximizing consumer benefits' (SGA, 2011) and 'the new energy consumer' (Accenture, 2011; Zpryme, 2011) reinforce and sustain the ABCD model by focusing on the human mind and its attitudes, opinions, drivers, values and choices. My intention is not to suggest that this entire collection of theories is completely wrong or invalid, but rather to acknowledge it as a unified compilation of concepts that dominates and pervades energy management and policy at the exclusion of others.

There is now a well-established critique of this dominant understanding of demand including its limitations in achieving long-lasting transformation in the ways we think about and use resources such as energy (Guy and Shove, 2000; Kempton and Montgomery, 1982; Lutzenhiser, 1993; Shove, 2003; Southerton et al., 2004; Wilhite et al., 2000; Wilk and Wilhite, 1985). Common critiques (and alternative framings) highlight the limited ability and willingness of consumers to make autonomous and rational decisions (Southerton et al., 2004), and the role that technologies and infrastructures play in mediating and co-shaping demand (Van Vliet et al., 2005).

One pervasive and problematic assumption underpinning the ABCD model is that demand managers must preserve, maintain and in some cases enhance existing lifestyles in accordance with individuals' attitudes, choices and values. In current energy policy and management, the emphasis is on reducing or shifting demand, 'without affecting the service provided' (NERA, 2008, p. 17), and 'in a way that avoids significant impacts on comfort and lifestyle' (Reidy, 2006, p. 4). The discourse is one where demand is shifted or shaved, while the services provided are maintained and sometimes enhanced. What is lacking is an understanding that practices, and the demand they create or require, are constantly changing, often in unpredictable and unexpected ways (Shove, 2003).

The idea of upholding existing ways of life is quite strange when we consider the scale and rate at which everyday practices are currently and constantly changing. For example, histories of domestic technologies (Schwartz Cowan, 1989) and comfort (Cooper, 1998) mark dramatic changes in what we consider to be normal cleaning, heating and cooling practices. In Australia, air-conditioning penetration has almost doubled in the last ten years to about 65 per cent, marking a dramatic change in how Australians keep cool in their homes (DEWHA, 2008). Similarly, there have been rapid changes in home computing and entertainment practices, resulting in an increase in appliances and gadgets such as laptops, smart phones, game consoles and home theatres (Harrington et al., 2006), The ABCD model masks the complexity of these changing socio-technical configurations of 'normality' across different timeframes, places and cultures (Shove, 2003). However, while there is now clear recognition that alternative paradigms and ways of understanding change are required, little attention has been paid to what this might mean for demand managers and the problem of peak electricity demand.

A SOCIAL PRACTICE PERSPECTIVE

There is no unifying agreement or definition of social practice theories, yet the list of issues to which theorists are now contributing is rich and diverse (Schatzki, 2001). This chapter refers to these theories interchangeably as 'practice theory', 'social practice theory' and a 'social practice perspective'. I am interested in the 'organization, reproduction, and transformation of social life' (Schatzki, 2001, p. 1) or the ways in which social practice theory constructs a distinctive social ontology, which differs markedly from that of the ABCD model. Table 2.1 helps to situate and clarify some of the distinctions between social practice and ABCD theories. It begins by

Table 2.1 Contrasting assumptions from ABCD and social practice theories

ABCD theories	Social practice theory
The world is populated by people	The world is populated by practices
People and their barriers, drivers, attitudes, values, opinions, choices and/or norms are the central unit of analysis and change	Practices (and their elements) are the central unit of analysis and change
Emphasis on changing people and their consumption/demand	Emphasis on the changing elements of practices
Technology, supply systems and people are separate from each other	Technologies and supply systems are elements of practices
People have agency	Practices, people and things have agency
People change through targeted information, education, price signals, social norms, community interaction etc.	Practices circulate and change through changing or mixing elements, and through innovation in practice
Change is orderly, predictable and controllable	Change is emergent, dynamic and often uncontrollable
Efficiency improvements and demand reductions are long-lasting	Practices are constantly changing along trajectories that may negate efficiency and conservation improvements

highlighting a simple yet significant premise: that the world is constructed and ordered by social practices, rather than individuals and their attitudes, behaviours or choices. In this sense, practice theory:

> Expresses itself decisively in a rejection of the modern conviction that mind is the central phenomenon in human life: the source of meaning, the receptacle of knowledge or truth, the wellspring of activity, and the co- or sole constitutor of reality. . . . Practices, in sum, displace mind as the central phenomenon in human life. (Schatzki, 2001, p. 11)

This does not mean that individuals become redundant or unnecessary, but rather that 'practices are the source and carrier of meaning, language, and normativity' (Schatzki, 2001, p. 12), or as Reckwitz (2002b, p. 254) argues: 'wants and emotions . . . do not belong to individuals but – in the form of knowledge – to practices' (Table 2.1). In this sense, beliefs, attitudes and values – the common building blocks of the ABCD model – can be thought of as arising from (and being cultivated within) practices rather than people. Individuals take on a very specific role in social practice theories as 'carriers' or 'performers', who are both 'captured' by practices and simultaneously constitute them through their reproduction of them (Shove and Pantzar, 2007, p. 156).

Schatzki (2002) distinguishes between a practice as being both a coordinated entity and a performance that is actualized and sustained through individuals' reproduction of it. The entity consists of a mix of inter-related elements, of which most practice theorists have their own definitions (see Gram-Hanssen, 2009 for a summary). For the purposes of this discussion, I identify these elements as being 'common understandings' about what the practice means and how it is valued, 'rules' about what procedures and protocols must be followed and adhered to, 'practical knowledge' about how to carry out and perform a practice and 'material infrastructure' – or the 'stuff' that makes the practice possible, sensible and desirable (Shove et al., 2007; Strengers, 2009). Elements are not deducible or always distinguishable – each should be viewed as interconnected and intersecting with each other (Pantzar and Shove, 2010).

For example, a practice theory perspective might view the increase in residential air-conditioning as the changing practice of household cooling, involving the complex co-evolution of material infrastructures (changing housing formats, central heating and cooling, the affordability and availability of the air-conditioner) (Strengers, 2010a); common understandings of air-conditioning as a normal and necessary service (Ackermann, 2002; Cooper, 1998), and changing notions of 'air', 'health' and 'wellbeing' associated with indoor climate and temperature (Shove, 2003); practical knowledge about how to cool the body and home (Strengers and Maller, 2011); and rules about how to use and install the air-conditioner (Kempton et al., 1992). This is distinct from accounts of the air-conditioner's rapid diffusion in western societies that privilege processes of market economics (affordability and availability), cultural symbolism (the air-conditioner's role as a 'status' object) or changing individual choices and lifestyle needs.

Table 2.1 also highlights the different role afforded to technologies and infrastructures in some social practice theories. In following a significant 'posthumanist' minority from the field of science and technology studies (Schatzki, 2001), this chapter positions technologies and infrastructures as key elements of practices: they take on the role of 'actants that can suggest and transform practices' (Hawkins and Race, 2011, p. 115). The inclusion of material infrastructure as an element of practices marks a significant departure from the supply–demand divide permeating the energy sector, in which technologies and human action are clearly separated and subject to different disciplinary approaches and models of change (for example, engineering versus psychology). In contrast, energy technologies and infrastructures 'necessarily participate in social practices just as human beings do' (Reckwitz, 2002a, p. 208).

Another critical distinction noted in Table 2.1 is the shift in focus from the consumption of resources to the practices in which all consumption is implicated (Warde 2005). In other words, people consume resources in order to carry out the day-to-day practices that they make possible (Warde, 2005). In this context, peak electricity demand policies and programmes are responding to changing social practices and, in particular, to the resource constraints and challenges these pose. This is not, in and of itself, a unique insight. The energy industry often talks about 'end-use services' as being the 'driver' of demand, recognizing that it is the activities that use energy that generate demand. However, where a practice theory perspective differs is that these end-use services are not viewed as the outcome of attitudes, behaviours or choices (ABC) but rather as the product of social practices.

In summary, social practice theory provides a distinct account of everyday life and social change that differs markedly from the ABCD model. The focus shifts from individual and autonomous agents, or self-directive and purposive technologies, and onto assemblages of common understandings, material infrastructures, practical knowledge and rules, which are reproduced through daily routines. It follows that the role of the demand manager is also repositioned. They can no longer be seen as purposive agents in the process of change, steering practices on particular courses. Their ability to affect change is 'complicated and qualified' by the 'emergent and uncontrollable trajectories' of practices (Shove and Walker,

2010, p. 475). Such qualifications necessarily reframe and potentially reject the role of traditional change agents in the reproduction and transformation of everyday life – requiring attention to what role they do, or might have, in reorienting and reordering it.

REFRAMING THE PEAK ELECTRICITY DEMAND PROBLEM

The role of demand managers depends, in part, on the ways in which the problems they seek to address are defined, and the strategies they employ to address them are established. Table 2.2 contrasts dominant ABCD questions used to frame the issue of peak electricity demand with those that might emerge from social practice theory. There is a fundamental shift in the units of enquiry and analysis: from resource consumption, technologies and individual behaviours to social practices, their elements and carriers. Alternative approaches are identified by examining how practices are changing and likely to change in the future, rather than extrapolating current resource consumption growth rates and projecting them forward (Table 2.2).

Rather than being viewed as a definitive framework, theory or method, these questions reposition the policy problem of peak electricity demand. Instead of asking how demand can be shaved or shifted, demand can be more usefully thought of as a symptom or by-product of changing social practices. The aim becomes one of understanding the elements and reproduction of problematic peaky practices, and attending to the dynamics of transformation and innovation that are occurring. The fifth section provides some insight into how and where these opportunities might arise, and how this repositions and broadens the role of demand managers.

REPOSITIONING THE ROLE OF DEMAND MANAGERS

Rethinking Provider–Consumer Relationships

As discussed earlier, electricity systems are currently managed within supply-demand siloes. This prioritises a provider–consumer relationship (Guy and Marvin, 2001), whereby the provider's role is to provide and maintain the supply of resources, and the consumer's role (and right) is to consume them. In the context of peak electricity demand, this is a problematic relationship, and one which legitimizes 'rights' to peaky practices

Table 2.2 Contrasting social theories applied to the issue of peak electricity demand

ABCD theories			Social practice theories		
Questions	Units of enquiry/ analysis	Approach	Questions	Units of enquiry/ analysis	Approach
Why is peak electricity demand rising?	Resource demand	Analyse market trends in appliance usage and sales	Why is peak electricity demand rising?	Practices	Identify which practices are contributing to changing demand; analyse their history and current trajectory
What are the projected demand trends into the future?	Resource demand, demographic segments	Analyse demand trends based on current projections (quantitative extrapolations)	How and why have relevant practices changed? How and why are they changing now?	Practices	Analyse how and why practices are changing
What are the barriers and drivers of demand? How can they be removed/ encouraged?	Barriers and drivers	Identify barriers and drivers; provide information, education and incentives to overcome/encourage them	What are the processes of practice change? How can they be supported?	Practices	Support and encourage innovation in practice; attempt to reorient practice elements
Which behaviours should be encouraged and what demand should be shifted?	Load profiles, social norms and values	Identify 'discretionary' and 'non-discretionary' demand; discretionary demand targeted	How are needs and wants constructed? How could they be constructed differently?	Practices, needs and expectations	Analyse how needs and wants are constructed through practices; identify and support processes of change

Table 2.2 (*continued*)

ABCD theories			Social practice theories		
Questions	Units of enquiry/ analysis	Approach	Questions	Units of enquiry/ analysis	Approach
How can people be educated to better manage their demand?	Resource management knowledge and information	Promote information and education that encourages household resource management	How can new practical knowledge, common understandings and rules be circulated?	Practices (practical knowledge, common understandings and rules)	Assist in the circulation of alternative practical knowledge, common understandings and rules
How can people be encouraged to make cost-reflective decisions about their consumption during peak times?	Individual behaviours and drivers; pricing signals	Develop cost-reflective pricing programmes or incentive schemes to shift or shed peak demand	How do common understandings and rules emerge from different pricing schemes?	Practices (common understandings and rules)	Identify pricing schemes that facilitate less peaky practices by reorienting the elements of these practices
How can demand be made more efficient?	Efficient and load-shifting technologies	Identify and promote efficient and 'smart' technologies	How are technologies and infrastructures co-shaping practices?	Practices	Identify and support technology and infrastructure that enables less peaky practices

(such as air-conditioning) or services (unwavering electricity demand) as non-negotiable needs (Shove and Chappells, 2001). In contrast, we can conceptualize the electricity system (supply) – as vast and intractable as it might seem – as an element of electricity-consuming social practices, informing what makes sense for householders to do during (and outside) peak periods.

Small-scale case studies of localized, community managed or distributed supply systems, such as solar panels (ATA, 2007) and micro grids (Chappells and Shove, 2004), provide examples of the dynamics involved when the supply–demand divide is collapsed and electricity systems become an active element in the practices they enable. Discussing micro grids in the UK, Chappells and Shove (2004, p. 139) suggest that:

> In these situations the distinction between provider and consumer collapses, opening up new opportunities for the coordination of demand and supply and for the 'real-time' management of resources and resource-consuming activities.

In this example, the electricity system realigns other practice elements and reorients current routines. In practice, this might involve householders' turning off unused or 'unnecessary' appliances, rescheduling 'discretionary' practices such as laundering to times when there is adequate demand or seeking alternative ways to undertake practices, such as putting on a jumper or going for a brisk walk in cold weather instead of turning the heater on. Such situations create opportunities for new needs to take hold around the variability of supply, similar to the ways in which consumers react and adapt to fuel restrictions.

Following this line of enquiry, Marvin and Perry (2005, p. 86) found that during the UK fuel crisis of 2000, network disruptions shaped 'innovative coping strategies that may have the potential for reshaping the user's relations with the network'. In their study of people's capacity to cope with this short-term crisis, they suggest that users adopt three main strategies: (1) they suspend their existing meetings or don't attend work; (2) they adjust the way they get to work by car sharing or altering other domestic routines; and (3) they adapt their normal routines significantly to cope with the crisis, such as working from home (Marvin and Perry, 2005).

In these examples, the role of the consumer changes: they are no longer passive recipients of complex networks and systems (electricity or fuel), but co-managers of their own practices, involving dynamics of both supply and demand (Chappells and Shove, 2004; Strengers, 2011a; Van Vliet et al., 2005). This is distinct from the expectation for constant and unwavering demand embedded into modern electricity systems, policies and regulations. Trentmann (2009, p. 80) highlights the problematic nature of this expectation, whereby consumers must be 'protected' from disruption, while they simultaneously 'play an active role in absorbing and coordinating them, in some cases even generating them'.

In Australia, this is most clearly illustrated in the example of residential cooling practices involving air-conditioning, which can lead to blackouts on hot summer days. Utilities are legally obligated to upgrade electricity infrastructure in attempts to meet this demand, despite the inefficiencies and costs it creates for networks and consumers (Strengers, 2010a). These interactions between supply and demand remain largely unaddressed in the policy and strategy siloes that dominate the energy sector. A practice theory perspective encourages us to collapse the separate roles and responsibilities of 'suppliers' and 'demanders', to critically examine how electricity systems uphold or challenge existing (problematic) needs, and how they can potentially enable innovative co-management opportunities.

Such attempts do not necessarily need to involve new material infrastructures such as micro grids, or 'crises' such as blackouts and fuel restrictions, but may relate to the rules or common understandings associated with existing supply systems. For example, variable pricing regimes, such as dynamic (or critical) peak pricing, or load limiting schemes whereby householders agree to cap their peak demand, can replicate the disruption and variability of large-scale electricity systems at the household scale. While these demand management strategies are primarily framed within a supply–demand paradigm, whereby suppliers must compensate and entice consumers to shift their demand, they could also be reconceptualized as a form of co-management.

For example, in Australia, dynamic peak pricing commonly refers to an electricity pricing scheme that charges 10–40 times the off-peak rate of electricity during peak 'events' that last for approximately four hours and are called up to 12 times a year (generally on very hot or cold days). This pricing strategy is designed to shift or shed demand during short peak periods to avoid power shortages resulting from peak electricity demand. The viability of this approach was evaluated in a small study of Australian households, who were found to rearrange their daily routines in response to this price signal, self-imposing their own blackouts or power cuts by undertaking alternative practices, shifting peaky practices to other times of the day, switching off power at the meter box or power point and/or leaving the house (Strengers, 2010a).

Similar to the studies discussed above (Chappells and Shove, 2004; Marvin and Perry, 2005), in this 'man-made' disruption, 'the temporal fragility of habits and the elasticity of everyday life' (Trentmann, 2009, p. 68) was revealed as householders demonstrated their creativity and practical skills in reorganizing and innovating daily routines. Here, the price signal took on the role of revealing the peakiness of the electricity system – resulting in non-discretionary practices, such as air-conditioning, being called into question. These findings support Trentmann's call for a more critical examination of the role of breakdowns and disruptions in managing everyday life – through pricing schemes, micro grid systems or load-limiting programmes (whereby the electricity load delivered to a home is controlled, varied or limited in line with the inherent fluctuations of the system). For demand managers, this might mean facilitating alternative relationship arrangements, or forms of co-management, which activate or accentuate the materiality of electricity systems as an element of social practices.

Working Beyond the Scale of Demand

Another practical strategy for demand managers is to refocus their attention beyond traditional boundaries and understandings of demand. In thinking about how practices are co-constructed, demand managers can identify a wider range of human and non-human stakeholders that are either deliberately or inadvertently involved in shaping and shifting the elements of peaky practices. Rather than seeing such issues beyond the scope of their role, a practice theory perspective necessitates viewing them as being integral to it.

For example, there are opportunities to think beyond the domain of the home in attempts to keep people cool. There may be scope to consider the establishment of 'cooling centres' or to promote and expand existing cool destinations, such as libraries, cinemas, pools, shopping centres and other community facilities, that shift peak cooling practices outside the home (Strengers and Maller, 2011). Similar to Australia's bushfire policy, one might imagine a 'stay or go' plan for households experiencing extreme heat, and the promotion of 'heat plans' and 'heat hotlines' for those at risk. This does not seem unreasonable when we consider that heat stress causes more deaths than floods, cyclones and bushfires (Maller and Strengers, 2011). Such activities potentially reorient the elements of peak cooling practices; for example, the material infrastructure of the home may no longer be considered essential to staying cool. This creates opportunities for other practice elements to adjust and realign in ways that potentially meet the objectives of demand managers.

Thinking within the domain of the home, changing housing designs and infrastructures have played a critical role in normalizing air-conditioning appliances to produce new practices of household cooling (Ackermann, 2002; Shove, 2003). In Australia, these changes have included the rise of open-plan living and central heating and cooling, combined with substantial increases in floor space and the decline in some thermal features such as eves (DEWHA, 2008; Strengers, 2010a; Wilkenfeld, 2004). In contrast, a house designed to prioritize passive thermal comfort, using appropriately placed windows, blinds and shading, potentially changes what makes sense for householders to do to achieve 'coolth' (Prins, 1992). Studies of adaptive thermal comfort support this proposition, finding that building occupants exposed to a range of 'passive' technologies and infrastructures tolerate and enjoy a wider band of temperatures than those living in climate-controlled situations (Brager et al., 2004; Nicol and Roaf, 2007). In contrast, policies and regulations that encourage, or do little to discourage, the trend towards central heating and cooling serve to legitimize and normalize a particular form of energy-intensive comfort (Chappells and Shove, 2005). De Dear and Brager (2002, p. 550) elaborate:

> People living year-round in air-conditioned spaces are quite likely to develop high expectations for homogeneity and cool temperatures, and may become quite critical if thermal conditions in their buildings deviate from the centre of the comfort zone they have come to expect. In contrast, people who live or work in naturally ventilated buildings where they are able to open windows, become used to thermal diversity that reflects local patterns of daily and seasonal climate variability.

These authors point to the ways in which housing infrastructures prioritize needs and expectations for particular types of cooling. In this way, we can see how houses, and the technologies and infrastructures which constitute them, can also be viewed as elements of practices that deserve further attention from demand managers.

However, while housing is often identified as a major 'contributor' or 'factor' to changing expectations of cooling, there is no way of addressing it within the ABCD model, apart from encouraging householders to buy more thermally efficient houses, or make relevant retrofit choices when they renovate their home. More worryingly, because peak electricity demand is primarily framed as an 'energy' issue rather than a housing one, such concerns are not addressed in housing policy – where relevant policies and regulations are formulated (Strengers and Maller, 2011). Equally problematically, professions that might influence the practices of household cooling, such as planners, designers, builders and developers, see these concerns as being largely beyond their role.

Without addressing these issues, the role of demand managers may be limited to one of advocacy or negotiation with the housing sector. Some demand managers have attempted this task with limited success. However, many more have not tried because they view this activity as being beyond the scope of their role (Strengers, 2010b). A crucial aspect of the problem is that those who are not tasked with the role of changing demand (such as the housing industry) have a substantial influence on its continuing transformation. Such issues serve to illustrate how the traditional siloes of energy policy and management inadvertently promoted by the supply–demand paradigm leave integral elements of practices overlooked or dismissed.

Promoting New Wants and Needs

A far more blatant approach for demand managers, or any self-identified change agent, is to actively debate and challenge taken for granted lifestyle 'needs'. Unlike advertisers and marketers, who study current practices in order to change or reinforce them, demand managers spend a lot of time and money asking consumers what their needs are, but make very few attempts to establish new ones. Indeed, in many cases they go out of their way to uphold assumed needs by, for example, encouraging pre-cooling prior to a dynamic peak pricing event, which reinforces the assumed need for air-conditioning during these times (see, for example, CountryEnergy, 2004).

Where attempts to instigate change are made, resource-centric information and awareness campaigns focus on 'saving' resources and money, yet fail to address expectations and needs – indeed in most cases they explicitly avoid them. For example, Sustainability Victoria's (SV, 2009) public media campaign encouraging householders to switch to cold water in the laundry in order to save money and energy overlooks the reasons why people use hot water in the first place, namely, to produce clean and hygienic clothing. Within this context, the energy sector's focus on appealing to householders' green motives, encouraging rational cost reflection and providing consumption feedback is often drowned out by the promotion of new expectations and aspirations associated with practices (such as the need to maintain higher levels of hygiene and cleanliness), and/or the commitment or nostalgic attachment to continuing domestic traditions (such as how someone was taught to do their laundry). The silence from demand managers on subjects considered a matter of 'personal choice' can be partly attributed to understandings of their roles.

For example, it is not normally considered the responsibility of demand managers to comment on the whiteness and brightness of laundry. Furthermore, such matters are considered 'private' and 'personal', rather than the domain or responsibility of utilities, NGOs and governments. In contrast, these issues are core business for marketing and advertising agencies (such as Datamonitor, 2008a, 2008b), who keep track of current trends and changing practices, providing advice to companies on where gaps exist for new or existing products. Marketing agencies attempt, with varying degrees of success, to insert new common understandings, material infrastructures and rules into practices, and provide practical knowledge about how to do them better (usually in ways that require specific products). In short, they promote practices (Shove and Pantzar, 2005), borrowing elements from others (such as common understandings of hygiene and cleanliness) to achieve their aims, in some cases counteracting the objectives of demand managers.

There is a strong need to critically reflect on the type of promotion demand managers currently carry out, and to shift from promoting energy and financial savings to promoting (and debating) practices. This means that, rather than trying to establish and embed new environmental, resource-oriented and economic understandings into every practice – a strategy that is likely to result in 'preaching to the converted' – demand managers should consider playing a more active role in shaping or counteracting the elements already associated with practices, or facilitate opportunities for new ones to form. Drawing again on the laundry example, this might involve attempting to counteract the assumption that cold water doesn't clean clothes as effectively as warm or hot water, an approach cold water detergent promoters are already taking (see, for example, Colgate, 2010). This is a matter of thinking about and attempting to shift the understandings embedded into a practice, rather than attempting to instil new environmental morals and economic responsibility into individuals.

In many ways, demand managers are already well aware of these strategies, although they are often applied to achieve the opposite objective; namely, to increase resource consumption. This is due to the simple reason that, notwithstanding considerable efforts to encourage energy conservation and efficiency, most energy utilities still make money by selling power (Strengers, 2010b). For example, the Australian retailer Origin Energy has a strong focus on environmental sustainability in all of its marketing materials, but is actively involved in promoting a range of energy appliances through its store, including air-conditioners (Origin, 2012). In promoting efficient air-conditioners to households that may or may not have one, utilities are effectively promoting air-conditioning and the cooling practices it enables (Shove, 2004).

The issue is slightly more complicated for electricity distributors, who have an economic incentive to shift peaky practices to other times of the day, but no real incentive to reduce them. An efficient distribution and transmission network is one where 'hot spots' of demand are moved into 'cold spots', creating an even, steady load of electricity at all times (Guy and Marvin, 1996). Using the example of household cooling, this would mean that distributors have the most to gain from encouraging pre- and post-cooling outside peak events, or indeed encouraging 24-hour climate-controlled environments where there is constant and consistent load, thereby increasing, reinforcing and potentially stabilizing the expectation for climate-controlled residential environments. Such issues suggest the need to pay attention not only to the peaky and resource-intensive practices of householders but also the practices of demand managers, and their agency to affect change.

ASSIGNING AGENCY AND IDENTIFYING AGENTS

This chapter began by identifying demand managers as change agents, implying that they have the agency to affect and direct change. Before concluding, it is worth clarifying how understandings of agency and agents necessarily differ between ABCD and social practice theories. In the ABCD model, agency is attributed to individuals, as self-interested and rational agents who have the power and choice to change their own actions – and to demand managers, who have the power to change people. In practice theory, this agency does not disintegrate, but rather is redirected and circulated through the reproduction of practices, where:

> . . . agents are body-minds who 'carry' and 'carry out' social practices. Thus, the social world is first and foremost populated by diverse social practices which are carried out by agents. (Reckwitz, 2002b, p. 256)

In this sense, two sets of practices become critical to any discussion concerning agency and change: the practices of demand management, which are carried out by demand managers, and the peaky practices that demand managers seek to change, which are primarily carried out by householders. Both sets of practices raise critical insights for assigning and understanding agency.

The first encourages attention to the practices involved in shifting and shedding demand. The delivery of strategies such as variable pricing, consumption feedback and direct load control can be conceptualized as historical situated practices, subject to their own 'pockets of stability and pathways of innovation' (Shove, 2010, p. 1278). Such strategies are invariably bound and contained by common understandings, practical knowledge, rules and material infrastructures defining what it means to manage demand and what responsibilities are assigned to the professions tasked with that role. While this chapter has not focused on these demand management practices, and the routes of change that might be open to reconfiguring them, this is a worthy topic for future investigation.

The second set of practices concerns those that demand managers seek to change; the everyday practices that are carried out by householders. Rather than ascribing demand managers with the agency to change these practices, we could reconceptualize these professions as potential manipulators of household practices – that is, people who (to varying degrees of success) are attempting to shift practice entities and contribute to new forms of circulation and reproduction. However, the successfulness of their attempts depends on the ways in which these changes are resisted or incorporated into the practice performances of householders. This redirects demand managers to attend to the more complicated and subtle role they (and others outside their professions) play in reorienting social practices and their elements, not necessarily deliberately or predictably.

This redirection of agency and agents is not merely a matter of semantics – it has quite profound impacts on what 'making change' means and how it happens. If an ABCD understanding of agency remains, there is a legitimate concern that the change agent professions will attempt to 'target' and 'drive' specific practices (as they currently target individuals) deemed 'good' or 'bad', 'discretionary' or 'non-discretionary'. Furthermore, they will overlook the important role of material infrastructures, which can also take the form of a change agent, reorienting other elements of practices as is the case with electricity systems, air-conditioners and housing infrastructures.

We might then conclude that the term 'change agent' is inherently unhelpful in moving beyond theories of demand: firstly, because social practice theory suggests that the change agent professions do not have the agency to affect change, at least not in the purposive sense commonly assumed; and secondly, because labelling one group of professions 'change agents' reinforces a siloed understanding of social change, where some professions can and should make change happen, whereas other potentially influential ones (like engineers) can't and shouldn't.

Alternatively, social practice theory points towards the identification and establishment of a new breed of change agents that extends beyond the demand management professions, and that is both deliberately and inadvertently reconfiguring the elements of problematic practices. Using the example of the changing practices of household cooling, these agents might include air-conditioning technologies and their designers, manufacturers and marketers; housing infrastructures and their developers, builders, planners and policy makers; and the engineers, regulators and policy makers involved in planning, designing, building and coordinating the large energy systems on which peaky (and non-peaky) practices now depend.

Of course, we must not forget the most critical change agents of all; the performers of household practices, those everyday innovators who, in the context of extreme heat, blackouts and changing cooling technologies, are variously involved in inventing and adapting ways to keep cool (Strengers and Maller, 2011). In contrast to demand management's current preoccupation with security supply and maintaining (or enhancing) current 'needs', these agents of everyday life demonstrate considerable adaptiveness and inventiveness in modifying, scheduling and transforming current routines when the elements of practices are reconfigured.

CONCLUSION

In the energy sector, where an ABCD model pervades attempts to shift and shed peak demand, social practice theory provides an alternative framing that significantly redefines the problem and potential responses to this internationally significant issue. From this perspective, the problem of peak demand can be usefully viewed as a symptom of changing expectations and conventions associated with everyday household practices, such as cooling, heating and entertaining. Wants, needs, values and expectations, which are commonly understood as originating from humans, emerge from and within practices. This perspective reorients the purpose and function of change agent professions such as demand managers.

This chapter has argued that going beyond the human mind allows for a more complex understanding of the ways in which changing elements of social practices construct and generate particular expectations for energy services and the practices they enable. This perspective potentially challenges the electricity industry's current preoccupation with securing the electricity grid, preserving current services and protecting consumers from their fluctuating demand – all of which serve to reinforce and potentially accelerate the very expectations that cause peak demand in the first place. Social practice theory can refocus attention on the changing elements of problematic peaky practices undertaken within the household, and on the human and non-human 'stakeholders' deliberately and inadvertently reorienting their trajectories.

The shift from people to practices challenges not only what the role of a change agent might be but who a change agent is. An engineer designing a new load management technology or upgrading the distribution network may be as important, if not more so, as the traditional change agent professions of behaviour change and demand management. Similarly, policy makers tasked with the role of achieving social and behavioural change are as relevant as those making policy concerning electricity or housing infrastructure. Social change can no longer be thought of as a confined process that takes place by manipulating and coercing human minds, but rather as a suite of transforming and intersecting social practices constituted by understandings, practical skills, rules and things. In this sense, directly or indirectly positioning certain professions as change agents may prove unhelpful given that such labels mask and potentially dismiss other processes and avenues of social transformation. Instead, an alternative breed of change agents needs to be identified and established that is involved, both deliberately and indirectly, in reconfiguring the elements of peaky and resource-intensive practices.

For those professions assigned the task of managing demand, this perspective may appear challenging – indeed it cuts to the core of what it means to 'be' a demand manager and 'do' demand management. Further work is required to understand how the practices of demand management are configured, what trajectories they are on and what opportunities there are to facilitate change within these professions. In the meantime, traditional change agents can use this perspective to reframe problems in a new light, innovate beyond the boundaries of their role and practise new forms of demand management that may send practices on less peaky trajectories. In accepting this challenge, they may bear witness to new and innovative forms of social change that are so urgently required.

ACKNOWLEDGEMENTS

Many thanks to members of the Centre for Urban Research's 'Beyond Behaviour Change' reading group (http://www.rmit.edu.au/cfd/beyond-behaviour) for fascinating discussions on this topic and suggestions for review; in particular Dr Cecily Maller, Helaine Stanley and Professor Ralph Horne. I am also grateful for the constructive feedback on previous drafts of this chapter provided by Professor Elizabeth Shove, Professor Gay Hawkins, Professor John Fien and Dr Zoe Sofoulis, as well as the insightful suggestions from two anonymous reviewers.

NOTE

1. This chapter was previously published in *Energy Policy*, **44**, 7614–23. It is reproduced here with kind permission from the publishers Taylor and Francis.

REFERENCES

Accenture (2011), *Revealing the Values of the New Energy Consumer: Accenture End-consumer Observatory on Electricity Management 2011*, Dublin: Accenture.

Ackermann, M. (2002), *Cool Comfort: America's Romance with Air-conditioning*, Washington, DC: Smithsonian Institution Press.

AEMC (Australian Energy Market Commission) (2011), 'Issues paper: Power of choice – giving consumers options in the way they use electricity', Sydney: Australian Energy Market Commission.

ATA (Alternative Energy Association) (2007), *The Solar Experience: PV System Owners' Survey*, Melbourne: Alternative Technology Association.

Brager, G., S. Paliaga and R.J. de Dear (2004), 'Operable windows, personal control, and occupant comfort', *ASHRAE Transactions*, **4695** (RP-1161), 17–35.

CEA (Consumer Electronics Association) (2011), 'Unlocking the potential of the smart grid – a regulatory framework for the consumer domain of smart grid', Arlington, VA: Consumer Electronics Association.

Chappells, H. and E. Shove (2004), 'Infrastructures, crises and the orchestration of demand', in D. Southerton, B. Van Vliet and H. Chappells (eds), *Sustainable Consumption: The Implications of Changing Infrastructures of Provision*, Edward Elgar: Cheltenham, UK and Northampton, MA, USA, pp. 130–43.

Chappells, H. and E. Shove (2005), 'Debating the future of comfort: environmental sustainability, energy consumption and the indoor environment', *Building Research and Information*, **33** (1), 32–40.

Colgate-Palmolive Company (2010), *Cold Power Advanced*, available at http://www.colgate.com.au/app/PDP/2xUltraConcentrate/AU/cold-power-2xultra-detergent-liquid-and-powder.cvsp (accessed 5 July 2010).

Cooper, G. (1998), *Air-conditioning America: Engineers and the Controlled Environment, 1900–1960*, Baltimore, MD: Johns Hopkins University Press.

CountryEnergy (2004), 'Here's everything you need to know to make an informed decision', The CountryEnergy Home Energy Efficiency Trial, CountryEnergy, Sydney.

Darby, S. (2010), 'Smart metering: what potential for householder engagement?', *Building Research and Information*, **38** (5), 442–57.

Datamonitor (2008a), *The Future of Home Hygiene and Clothing Care Occasions*, available at http://www.datamonitor.com/consumer (accessed 3 March 2009).

Datamonitor (2008b), *Sterilized Society: Consumer Attitudes Towards Hygiene and Cleanliness*, available at http://www.datamonitor.com/consumer (accessed 3 March 2009).

De Dear, R.J. and G.S. Brager (2002), 'Thermal comfort in naturally ventilated buildings: revisions to ASHRAE Standard 55', *Energy and Buildings*, **34**, 549–61.

DEWHA (Department of the Environment, Water, Heritage and the Arts) (2008), *Energy Use in the Australian Residential Sector 1986–2020*, Canberra: Australian Government: Department of the Environment, Water, Heritage and the Arts.

ETSA (2007), *Demand Management Program: Interim Report No. 1*, Adelaide: ETSA Utilities.

Gram-Hanssen, K. (2009), 'Standby consumption in households analyzed with a practice theory approach', *Research and Analysis*, **14** (1), 150–65.

Guy, S. and S. Marvin (1996), 'Transforming urban infrastructure provision — the emerging logic of demand side management', *Policy Studies*, **17** (2), 137–47.

Guy, S. and S. Marvin (2001), 'Urban environmental flows: towards a new way of seeing', in S. Guy, S. Marvin and T. Moss (eds), *Urban Infrastructures in Transition: Networks, Buildings, Plans*, London: Earthscan Publications, pp. 22–40.

Guy, S. and E. Shove (2000), *A Sociology of Energy, Buildings and the Environment*, London: Routledge.
Harrington, L., K. Jones and B. Harrison (2006), 'Trends in television energy use: where it is and where it's going', Paper presented at the 2006 ACEEE Summer Study on Energy Efficiency in Buildings, Asilomar Conference Centre, Pacific Grove, California, 13–18 August.
Hawkins, G. and K. Race (2011), 'Bottle water practices: reconfiguring drinking in Bangkok households', in R. Lane and A. Gorman-Murray (eds), *Material Geographies of Household Sustainability*, Farnham: Ashgate, pp. 113–24.
Herter, K., (2007), 'Residential implementation of critical-peak pricing of electricity', *Energy Policy*, **35**, 2121–30.
Hobson, K. (2006), 'Bins, bulbs, and shower timers: on the "techno-ethics" of sustainable living', *Ethics, Place and Environment*, **9** (3), 317–36.
Kempton, W. and L. Montgomery (1982), 'Folk quantification of energy', *Energy*, **7** (10), 817–27.
Kempton, W., D. Feuermann and A.E. McGarity (1992), '"I always turn it on super": user decisions about when and how to operate room air conditioners', *Energy and Buildings*, **18**, 177–91.
Lutzenhiser, L. (1993), 'Social and behavioral aspects of energy use', *Annual Review of Energy and the Environment*, **18**, 247–89.
Maller, C.J. and Y. Strengers (2011), 'Housing, heat stress and health in a changing climate: promoting the adaptive capacity of vulnerable households, a suggested way forward', *Health Promotion International*, **26** (1), 100–8.
Marvin, S. and B. Perry (2005), 'When networks are destabilized: user innovation and the UK fuel crises', in O. Coutard, R.E. Hanley and R. Zimmerman (eds), *Sustaining Urban Networks: The Social Diffusion of Large Technical Systems*, London: Routledge, pp. 86–100.
NERA (2008), *Cost Benefit Analysis of Smart Metering and Direct Load Control. Work Stream 4: Consumer Impacts*, Phase 2 Consultation Report, NERA Economic Consulting, prepared for the Ministerial Council on Energy Smart Meter Working Group, Sydney.
Nicol, F. and S. Roaf (2007), 'Progress on passive cooling: adaptive thermal comfort and passive architecture', in M. Santamouris (ed.), *Advances in Passive Cooling*, London: Earthscan, pp. 1–29.
Origin Energy Ltd (2012), *Origin*, available at http://www.originenergy.com.au/ (accessed 18 January 2012).
Pantzar, M. and E. Shove (2010), 'Understanding innovation in practice: a discussion of the production and reproduction of Nordic walking', *Technology Analysis and Strategic Management*, **22** (4), 447–61.
Pears, A. (2004), 'Megawatts or negawatts – distributed and demand-side alternatives to new generation requirements', Paper presented at the Australian Institute of Energy Symposium, Energy in NSW 2004: A Strategic Review, Australian Technology Park, Redfern, Australia, 20 May.
Prins, G. (1992), 'On condis and coolth', *Energy and Buildings*, **18**, 251–8.
Reckwitz, A. (2002a), 'The status of the "material" in theories of culture. From "social structure" to "artefacts"', *Journal for the Theory of Social Behaviour*, **32** (2), 195–217.
Reckwitz, A. (2002b), 'Toward a theory of social practices: a development in culturalist theorizing', *Journal of Social Theory*, **5** (2), 243–63.
Reidy, C. (2006), *Interval Meter Technology Trials and Pricing Experiments: Issues*

for Small Consumers, Sydney: Institute of Sustainable Futures, prepared for Consumer Utilities Advocacy Centre.

Schatzki, T.R. (2001), 'Practice theory', in T.R. Schatzki, K. Knorr Cetina and E. Von Savigny (eds), *The Practice Turn in Contemporary Theory*, New York: Routledge.

Schatzki, T.R. (2002), *The Site of the Social: A Philosophical Account of the Constitution of Social Life and Change*, Pennsylvania, PA: Pennsylvania State University Press, pp. 1–14.

Schwartz Cowan, R. (1989), *More Work for Mother: The Ironies of Household Technology from the Open Hearth to the Microwave*, London: Free Association Books.

SGA (2011), Maximising consumer benefits, Melbourne: Smart Grid Australia.

Shove, E. (2003), *Comfort, Cleanliness and Convenience: The Social Organisation of Normality*, Oxford: Berg Publishers.

Shove, E. (2004), 'Efficiency and consumption: technology and practice', *Energy and Environment*, **15** (6), 1053–65.

Shove, E. (2010), 'Beyond the ABC: climate change policy and theories of social change', *Environment and Planning A*, **42**, 1273–85.

Shove, E. (2011), 'Commentary: On the different between chalk and cheese – a response to Whitmarsh et al's comments on "Beyond the ABC: climate change policy and theories of social change"', *Environment and Planning A*, **43**, 262–4.

Shove, E. and H. Chappells (2001), 'Ordinary consumption and extraordinary relationships: utilities and their users', in J. Gronow and A. Warde (eds), *Ordinary Consumption*, London: Routledge, pp. 45–58.

Shove, E. and M. Pantzar (2005), 'Consumers, producers and practices: understanding the invention and reinvention of Nordic walking', *Journal of Consumer Culture*, **5** (1), 43–64.

Shove, E. and M. Pantzar (2007), 'Recruitment and reproduction: the carriers of digital photography and floorball', *Human Affairs*, **17**, 154–67.

Shove, E. and G. Walker (2010), 'Governing transitions in the sustainability of everyday life', *Research Policy*, **39**, 471–6.

Shove, E., M. Watson, M. Hand and J. Ingram (2007), *The Design of Everyday Life*, Oxford: Berg.

Sofoulis, Z. (2011), 'Skirting complexity: the retarding quest for the average water user', *Journal of Media and Cultural Studies*, **25** (6), 795–810.

Southerton, D., A. Warde and M. Hand (2004), 'The limited autonomy of the consumer: implications for sustainable consumption', in D. Southerton, B. Van Vliet and H. Chappells (eds), *Sustainable Consumption: The Implications of Changing Infrastructures of Provision*, Cheltenham, UK and Northampton, MA, USA: Edward Elgar, pp. 32–48.

Strengers, Y. (2009), 'Bridging the divide between resource management and everyday life: smart metering, comfort and cleanliness', PhD thesis, RMIT University, Melbourne.

Strengers, Y. (2010a), 'Air-conditioning Australian households: a trial of dynamic peak pricing', *Energy Policy*, **38** (11), 7312–22.

Strengers, Y. (2010b), 'Comfort expectations: the impact of demand-management strategies in Australia', in E. Shove, H. Chappells and L. Lutzenhiser (eds), *Comfort in a Lower Carbon Society*, New York: Routledge, pp. 77–88.

Strengers, Y. (2011a), 'Beyond demand management: co-managing energy and water consumption in Australian households', *Policy Studies*, **32** (1), 35–58.

Strengers, Y. (2011b), 'Designing eco-feedback systems for everyday life', *Proceedings of the 2011 Annual Conference on Human Factors in Computing Systems*, Vancouver ACM, Vancouver, pp.2135–44.
Strengers, Y. and C. Maller (2011), 'Integrating health, housing and energy policies: the social practices of cooling', *Building Research and Information*, **39** (2), 154–68.
SV (Sustainability Victoria) (2009), 'You have the power. Save energy', available at http://www.saveenergy.vic.gov.au/ (accessed 12 June 2009).
Trentmann, F. (2009), 'Disruption is normal: blackouts, breakdowns and the elasticity of everyday life', in E. Shove, F. Trentmann and R.R. Wilk (eds), *Time, Consumption and Everyday Life: Practice, Materiality and Culture*, Oxford: Berg, pp. 67–84.
Van Vliet, B., H. Chappells and E. Shove (2005), *Infrastructures of Consumption: Environmental Innovation in the Utilities Industries*, London: Earthscan.
Warde, A. (2005), 'Consumption and theories of practice', *Journal of Consumer Culture*, **5** (2), 131–53.
Wilhite, H., E. Shove, L. Lutzenhiser and W. Kempton (2000), 'The legacy of twenty years of energy demand management: we know more about individual behaviour but next to nothing about demand', in E. Jochem, J. Sathaya and D. Bouille (eds), *Society, Behaviour and Climate Change Mitigation*, Dordrecht: Kluwer Academic Publishers, pp. 109–26.
Wilk, R.R. and H. Wilhite (1985), 'Why don't people weatherize their homes? An ethnographic solution', *Energy*, **10** (5), 621–9.
Wilkenfeld, G. (2004), *A National Demand Management Strategy for Small Airconditioners: The Role of The National Appliance and Equipment Energy Efficiency Program (NAEEEP)*', Sydney: George Wilkenfeld and Associates for the National Appliance and Equipment Energy Efficiency Committee (NAEEEC) and the Australian Greenhouse Office.
Zpryme (2011), *The New Energy Consumer*, Austin, TX: Zpryme Smart Grid Insights.

3. Economic and non-economic factors driving household expenditure: methodological reflections on an econometric analysis

Scott Milne

INTRODUCTION

As part of the Economic and Social Research Council (ESRC) Research group on Lifestyles, Values and Environment (RESOLVE),[1] an econometric analysis of UK household expenditure was carried out. The analysis estimated the relative contribution of income, price and 'other effects' driving expenditure, without seeking to identify much less quantify those other effects. In this chapter, a series of methodological issues are considered in relation to the original study and to any subsequent enquiry regarding the underlying factors driving household expenditure.

To frame the discussion, the theory of rational choice and its key criticisms are outlined. Those criticisms include concerns about the boundedness of actual decision making, the role of habits and emotions in impeding rationality and the conflict between the notion of self-interest and apparently altruistic behaviour. The insistence upon the individual as the unit of analysis has also been challenged, and alternative approaches that give consideration to the role of social structures are discussed.

To explain how the rational individual came to dominate in mainstream economics, a short history of economic methodology is provided. While the classical economists treated the basic postulates of economics as a priori truths, the influence of logical positivism and falsificationism gave rise to concerns about the testability of assumptions. Those concerns were soon brushed aside, however, with the emergence of an instrumentalist approach for which assumptions were only required to be predictively useful, rather than realistic. Although this disregard for the underlying structures of the social world has been heavily criticized by economic methodologists, applied economics has proceeded in a highly mathematized and axiomatic manner ever since.

Next, the aforementioned econometric analysis is described. While the original study offered provisional policy implications, these are seen as problematic in the light of further considerations, illustrated in this chapter through a hypothetical example. It is argued that confidence in the effect of economic interventions is undermined when the 'other effects' besides income and price are left unexplored.

An instrumentalist approach might be to disregard such concerns, as long as the aggregate relationships between economic variables endure. However, it is suggested here that the very presence of 'other effects' besides income and price ought to give rise to consideration of the underlying structures influencing consumer behaviour.

The scientific realist perspective is proposed as a suitable philosophical orientation to support any subsequent enquiry, and while the chief proponent of realism in economics has been sceptical of the value of econometrics, other commentators have suggested a complementary role for statistical analysis alongside more qualitative approaches. A series of possible avenues for future work are therefore suggested.

RATIONAL CHOICE

The core theory with regard to conceptualizing and explaining consumer behaviour in mainstream economics is that of rational choice (Hausman, 2008, section 5). According to rational choice theory, in any decision-making process, a person will assess the expected benefits and costs of the various options in a given situation, and will choose the option that will maximize individual gain, while minimizing losses. Therefore, a consumer with particular tastes and preferences, faced with a given level of income and a particular range of goods (at particular prices), will choose from those goods in such a way as to maximize their expected utility.

In rational choice theory, consumer tastes and preferences are treated as exogenous to the model and – in the strict version at least – any potential barriers to market information (about available goods and their prices) are ignored. As a result, the main point of intervention for policy makers seeking to facilitate behaviour change is the price mechanism. By changing the relative costs and benefits of different goods, for example, through taxes and subsidies, utility-maximizing consumers will change their behaviour accordingly.

The core assumption of rational choice has been heavily criticized, both from within the economics discipline and beyond. A good summary of some of the key criticisms is provided by Jackson (2005), for example. These criticisms are not exclusive to any one alternative perspective, and many critiques – internal and external to the discipline – have begun with the same observations and arrived at very different responses. Nevertheless, even from within the discipline, the limitations of the links made between rationality, choice and behaviour are clearly highlighted.

Bounded Rationality

Simon (1957), for example, suggests that in real life situations, obtaining information for decision making not only imposes a time cost but is often hampered by uncertainties regarding future outcomes. For this reason, he proposed a model of 'bounded rationality' in which, rather than reviewing all information to obtain the optimal outcome, individuals instead make do with a satisfactory outcome that meets a minimum level of utility.

Bounded rationality implies the use of heuristics in place of deliberative decision making. Viewed as time saving devices, heuristics might be satisfactorily incorporated into a somewhat weakened version of rational choice theory. However, unconscious or habitual behaviour would seem to contradict rational choice theory altogether in cases where it acts as a substantial impediment to a consciously preferred alternative.

Emotion and Choice

Further criticisms along these lines point to the significant role played by emotions in decision making. Again, allowing for an affective component in decision making need not contradict rational choice if the satisfaction of emotional needs were treated as simply another aspect of utility. Things get more complicated though when cognitive processes are seen as being fundamentally driven and shaped by emotions, which are themselves partly triggered by physiological factors. In this view, 'reason itself [is] a construct of our emotional responses to situations' (Jackson, 2005, p. 37). Not all criticisms of rational choice theory treat emotion in this way. Nevertheless, rational choice theory clearly lacks any explicit explanation of its own with regards to affective aspects of behaviour.

Altruistic Behaviour

Another strand of criticism relates to the individual as purely self-interested and points to the existence of altruistic behaviour. Rational choice theorists may seek to account for altruistic behaviour by expanding the notion of self-interest to include the family/tribe/species as a whole. Alternatively, seemingly altruistic behaviour may be put down to a simple 'feel good' factor, or the expectation that such behaviour will be reciprocated. But critics insist this is a move too far from the individual self-interest implied by the theory. Instead of developing the theory in a way that may provide scope for further investigation, these responses simply redefine self-interest so broadly that, in the end, the theory lacks all explanatory power.

Methodological Individualism

Another criticism of rational choice theory arises from a more general resistance to reductionism in the understanding of social structures. Methodological individualism – the idea that the individual constitutes the appropriate unit of analysis in the study of behaviour – was introduced to the social sciences by Weber [1978] [1956]; see also Heath, 2011). Weber argued that 'social collectivities' such as states and corporations 'must be treated as *solely* the resultants and modes of organization of the particular acts of individual persons' (Weber, 1978, p. 13, emphasis in original).

Critics of methodological individualism argue that there are simply too many factors relating to social structures that fail to correspond to the self-interest of individuals. This has led to the development of alternative accounts of social action wherein the structures themselves, rather than the individual, are treated as the main unit of analysis. The structuralist approach in turn has been accused by some of going too far. In defence of methodological individualism, Elster (1989, p. 158) insists that 'talk about institutions is just shorthand for talk about individuals who interact with one another . . . whatever the outcome of the interaction, it must be explained in terms of the motives and the opportunities of these individuals'.

Others have attempted to find a middle ground between individualism and structuralism. The approach termed 'methodological localism' (Little, 2009) marks an enhancement upon individualism in two ways. Firstly, localism 'affirms that there are large social structures and facts that influence social outcomes'. Secondly, while these structures are deemed to be possible only by virtue of being micro-founded in the 'actions and states' of individuals, those individuals are to be understood as 'socially constructed and socially situated . . . within a set of local social relationships, institutions, norms, and rules'.

To the extent that rational choice theory implies a strict form of methodological individualism, Jackson (2005, p. 39) concludes that there is a sufficiently strong case to be made against its 'under-socialization' of human behaviour.

In the face of such criticisms, it is reasonable to wonder how the rational individual came to play such a prominent part in contemporary economic theory, and whether there is some scope for a loosening of this axiomatic assumption, particularly in the pursuit of a more interdisciplinary enquiry into consumer behaviour.

A BRIEF HISTORY OF ECONOMIC METHODOLOGY

Despite their important contributions to economic thought, many of the early political economists neglected to address issues of methodology directly (Blaug, 1992 p. 51). Thus, the works of Adam Smith, David Ricardo and Thomas Malthus offer no explicit treatment of their philosophical approach to economic theorizing. Instead, Blaug points to later figures including Nassau William Senior (1836) and John Stuart Mill (1844, 1848) as the classical economists who did the most to develop the methodology of the young discipline (Blaug, 1992, pp. 51–2).

Basic Postulates as a Priori Truths

These early methodologists focused primarily on establishing the premises of economic theory, the basic postulates that would act as the building blocks of further reasoning. For Senior, these premises consisted of a priori facts that were 'the result of observation, or consciousness, and scarcely requiring proof, or even formal statement, which almost every man, as soon as he hears them, admits as familiar to his thoughts' (Senior, 1836, p. 129). Chief among these premises was the principle that 'every man desires to obtain additional wealth with as little sacrifice as possible' (Senior, 1836, p. 139).

These basic postulates were insufficient to deduce economic outcomes in the real world, however, due to the presence of 'disturbing causes'. Only in the absence – or complete specification – of disturbing causes could the implications of economic theory be expected to offer predictive power. Consequently, where the implications of a theory did not match real world experience, this predictive failure could be attributed to the presence of not-yet-specified disturbing causes rather than the falsehood of the basic postulates.

The methodological approach espoused by Senior, Mill and their contemporaries was endorsed as late as the 1930s (Robbins, 1932). For Robbins, there are a priori economic truths, general laws that form the basis of economic theory. While these must be supplemented with subsidiary postulates describing the historico-relative context in which we wish to apply them (market conditions, legal frameworks and so on), the laws themselves are not reducible to historico-relative features (Robbins, 1932, pp. 79–80).

Logical Positivism and Falsification

This view of economic laws as a priori truths, or in Robbins's words 'the stuff of our everyday experience' (Robbins, 1932, p. 79), came under increasing scrutiny, particularly with the emergence of logical positivism. Terence Hutchison (1938) set out to demarcate modern economic science from what he perceived to be the pseudo-scientific approach of previous economic methodologists. Hutchison invoked the logical positivism of the Vienna Circle, arguing that what distinguishes scientific from non-scientific statements is their testability. In an attack on the apriorism of earlier economists, he argued that laws in economics, as in other scientific disciplines, should be '*conceivably falsifiable*, though not *practically falsified*, empirically' (Hutchison, 1938 p. 62, emphasis in original).

According to Caldwell (2003 p. 115), Hutchison's commitment to positivist principles came as confirmation of changes already afoot within the discipline, and their widespread approval among economists coincided with the 'mathematization and quantification of economic theory' in subsequent decades. The extreme positivism of the 1930s matured into a more considered and comprehensive system by the 1950s, driven by concerns about 'falsifiability' and the nature of the testable propositions that were to replace a priori truths (Caldwell, 2003 p. 19). Attention to such matters fed through into economic methodology, culminating in Samuelson (1947) describing one of the central aims of economics as being to derive '[hypotheses] about empirical data which could conceivably be refuted if only under ideal conditions' (Samuelson, 1947 p. 4).

Friedman's Instrumentalism

But just as falsificationism was establishing itself within economics, Milton Friedman (1953) in an essay on economic methodology dismissed the very idea that economic theorists should proceed by testing assumptions. Friedman argued that the validity of a theory does not rest on the conformity of its assumptions with the facts: 'A hypothesis is important if it *explains* much by little, that is, if it abstracts the common and crucial elements from the mass of complex and detailed circumstances surrounding the phenomena . . . to be important, therefore, . . . a hypothesis must be descriptively false in its assumptions' (Friedman, 1953, p. 14 emphasis in original). Instead, he argued, the validity of a theory depends solely on whether it delivers accurate predictions; an approach known as instrumentalism.

Despite the essay coming under sustained attack, Friedman never saw fit to respond to his critics (Caldwell, 2003, p. 173). As a result, ambiguities have plagued the secondary literature around the essay, not least of all Friedman's meaning of 'realistic', a term that Caldwell notes has been variously interpreted as meaning 'testable, highly confirmed, or true' (Caldwell, 2003, p. 178).

Blaug (1992, p. 94) contends that Friedman's essay also lacks a proper treatment of the different kinds of assumptions that may be employed in economic theories, that is, assumptions about motivations, behaviour, functional relationships, boundary conditions and more. Without an underlying structure of assumptions, Blaug concludes, there is nothing that can be adjusted and improved if a theory initially fails in its predictions: 'It is for this reason that scientists usually do worry when the assumptions of their theories are blatantly unrealistic' (Blaug, 1992, p. 99).

Despite the array of methodological criticisms it encountered, Friedman's essay nevertheless persuaded a generation of economists that theories were ultimately to be judged according to the accuracy of their predictions (Blaug, 1992, p. 110) and that the alternative approach focusing on the testing of assumptions was 'fundamentally wrong' (Friedman, 1953, p. 14). As a result, Blaug argues:

> The prevailing methodological mood is . . . ultrapermissive within the limits of the 'rules of the game': almost any model will do provided it is rigorously formulated, elegantly constructed, and promising of potential relevance of real-world situations . . . . Modern economists frequently preach falsificationism . . . but they rarely practice it. (Blaug, 1992, p. 111)

THE RISE OF HOMO ECONOMICUS

The basic concept of rationality has endured throughout this long transition in economic methodology, although the emphasis on how that concept should be employed has shifted. Senior's core principle of rationality, namely, that 'every man desires to obtain additional wealth with as little sacrifice as possible' (Senior, 1836, p. 139), was subsequently elaborated on by Mill, who insisted that this principle should be treated as a mere abstraction, but one that was necessary for the science of political economy. Mill insisted that '[No] political economist was ever so absurd as to suppose that mankind are really thus constituted, but . . . this is the mode in which science must necessarily proceed' (Mill, 1844, p. 139).

Furthermore, Mill was adamant that there are many aspects of human conduct where 'wealth is not even the principle object' and to which economics does not apply (Mill, 1844, pp. 139–40). Instead, Mill argued that economics applies only to those areas of human conduct where the accumulation of wealth is the main objective and – for effective analysis – this main objective must be treated as if it were the sole objective in order to arrive at an approximation of reality. Crucially, though, that approximation must be adjusted to take account of non-economic impulses. Mill delivers a clear statement of the necessity of such accommodation:

> So far as it is known, or may be presumed, that the conduct of mankind in the pursuit of wealth is under the collateral influence of any other of the properties of our nature than the desire of obtaining the greatest quantity of wealth with the least labour and self-denial, the conclusions of Political Economy will so far fail of being applicable to the explanation or prediction of real events, until these are modified by a correct allowance for the degree of influence exercised by the other cause. (Mill, 1844, p.140)

Despite Mill's clarity in describing rational behaviour as an abstraction with limited applicability and his call for corrections to be made for other effects, the concept of 'homo economicus' was born in response to his work (Blaug, 1992, p. 55–6, p. 74). The view that rational behaviour could be employed directly as a basis of economic theory played into the wider mathematization and quantification of the discipline described by Caldwell (2003, p. 115).

Revisiting Rationality and Its Disturbing Causes

Although rational choice has come to dominate modern economics as an approximation of actual human behaviour, as opposed to a mere abstraction of certain aspects of it, Mill's writing makes it clear that the consideration of non-economic factors is embedded in the heritage, at least, of modern economic theory. Thus, any attempt to consider the role of non-economic factors in the study of consumer behaviour should not be seen as an attack on the discipline of economics but rather as an attempt to provide the kind of correction advocated by one of the fathers of political economy.

The following section discusses the results and implications of an econometric analysis that sought to explore such questions by estimating the role of economic and non-economic factors influencing UK household expenditure. The analysis raises some important questions regarding the conduct of research into consumer behaviour, questions that are explored with reference to the methodological issues discussed above.

ECONOMETRIC ANALYSIS OF THE DRIVERS OF EXPENDITURE

As part of the multidisciplinary Research group on Lifestyles, Values and Environment (RESOLVE), a team of econometricians set out to examine the key drivers of UK household expenditure. In the wider setting of RESOLVE, the goal was to understand how policy intervention might help to deliver a reduction in carbon emissions associated with household expenditure. The result was an econometric model called Econometric Lifestyle Environmental Scenario Analysis (ELESA) (Chitnis and Hunt, 2009a, 2009b, 2010, 2011).

Applying a Structural Time Series Model (STSM) (Harvey, 1991), household expenditure functions were estimated for 16 categories of goods and services[2] using aggregate time series data for the UK from 1964 to 2009. For each category, ELESA estimated the contribution of historical changes in prices and incomes to changes in household expenditure[3]. Crucially, the STSM approach also allows for the estimation of any stochastic underlying trends, that is, changes in the expenditure time series not attributable to price and income effects. The underlying trend in each category therefore represents the contribution of disturbing causes or (in the language of the STSM approach) 'unobserved components' to changes in expenditure.

In terms of results, the expenditure functions estimated in ELESA show that an underlying trend was apparent for each of the 16 categories of goods and services. In other words, there were no categories in which the economic factors of income and price were sufficient to fully account for expenditure changes. Nevertheless, it was possible to identify categories where income and price were stronger factors (relative to the unobserved components) and other categories where the disturbances were stronger (for a comprehensive discussion of the estimated contributions in each category see Chitnis and Hunt, 2009a, 2009b, 2010, 2011).

One of the goals of ELESA was to provide an evidence base to policy makers of the types of interventions that might be successful across different categories. With that in mind, the researchers concluded that economic interventions could be targeted at categories of goods and services where economic factors are relatively strong drivers[4]. Meanwhile, where unobserved components play a stronger role, economic interventions might be expected to have more limited impact and a mixture of economic and non-economic intervention may be required.

Although at a superficial level this seems like reasonable advice, upon reflection it raises important methodological questions regarding the nature of the unobserved components influencing expenditure. While it is one thing to estimate the net contribution of unobserved components historically, it is quite another to predict how those components will behave in the future, particularly when subject to intervention.

Unobserved Components Driving Expenditure

For the economic variables examined in ELESA, it is possible to identify precisely what these are and where they came from. The income variable is of course the aggregate of all household income in the UK. For an individual household, income may go up, down or remain unchanged, but the aggregate figure by definition gives the overall change in UK household income from one period to the next. Aggregate household expenditure works on the same principle: expenditure at the level of individual categories is simply the aggregate of spending on the particular goods and services that comprise that category (which may go up, down or remain unchanged from one period to the next). Finally, the price index series for each expenditure category is an implicit weighted average of all the price changes of the different goods and services within the category.

By contrast with those economic factors, the underlying trend is not aggregated up from its (unobserved) component parts, it is simply the estimated effect of 'everything else'. Although a series of non-economic factors like demographics, legislation, technical progress or tastes and preferences have been offered as possible components, in no sense were these quantified in the original study by Chitnis and Hunt. Of course, any analysis of time series data has to begin somewhere, and distinguishing an underlying trend from other variations is a useful first step.

In fact, in the most basic time series model introduced by Harvey, a function can be estimated in terms of: trend, cyclical, seasonal and irregular (error term) variations only, that is, without reference to any other variables than the time series itself (Harvey, 1991). Applying such a model in the case of household expenditure would mean that the impact of income and price remained hidden among the unobserved components contributing to the underlying trend. Against this approach, the model applied in ELESA – by incorporating income and price as determining variables – is already a step towards identifying some of the (otherwise) unobserved components.

The endogenization of these two economic factors by Chitnis and Hunt also helps to explain why those authors refer to the remaining unobserved components as 'exogenous non-economic factors'. The term is perhaps unfortunate though, since there could well be further economic variables among the unobserved components driving changes in household expenditure, such as interest rates, household wealth or taxation.

Clearly there are limits to how far these components can be endogenized into the model, but it is precisely the 'black box' nature of the unobserved components that leads to fundamental uncertainty in predicting the impact of any intervention.

Inside the Black Box of Unobserved Components

Consider two hypothetical categories of goods and services (A and B) for which the various contributions to expenditure changes over a given period are estimated to be identical: 45 per cent due to price changes, 45 per cent due to income changes and 10 per cent due to unobserved components. According to the policy conclusion drawn in the initial study, these both seem like suitable categories for some form of economic intervention. However, in the case of category A, the contribution of unobserved components is small because there is genuinely very little 'non-economic' disturbance upon the rational behaviour of consumers.

Meanwhile in the case of category B, where the net contribution of unobserved components is similarly small, this is in fact due to the near cancelling out of two sets of highly significant exogenous effects.

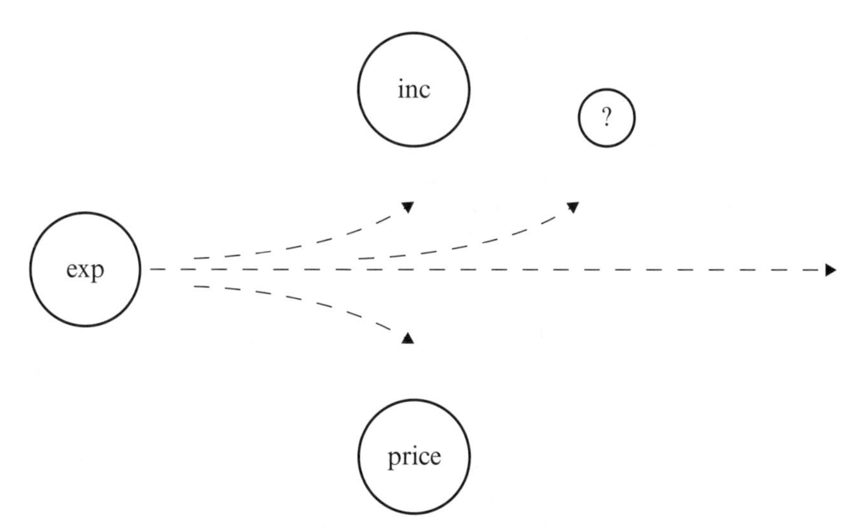

Figure 3.1 Category A where income and price are more significant drivers than the unobserved components. In this illustration (adopting the analogy of gravitation), income and price act in opposing directions, while the small contribution from unobserved components is entirely positive

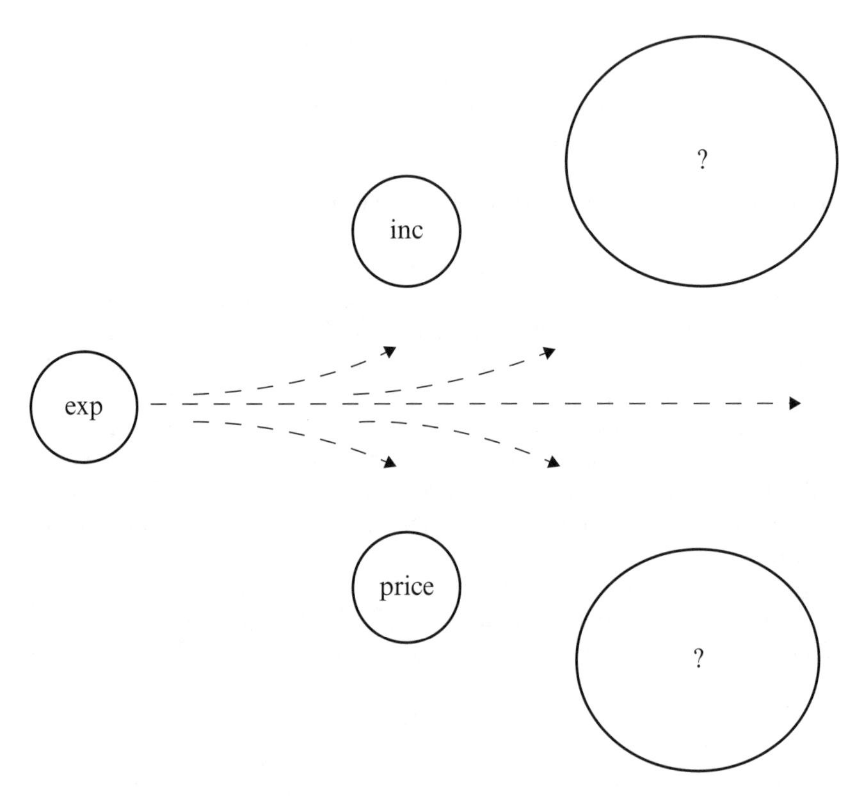

Figure 3.2 Category B where income and price effects are dwarfed by two sets of unobserved components. These components are opposed, however, such that ELESA would estimate a small net positive contribution, as in Category A

Regardless of what those exogenous effects actually are (for example, changes in demography, technology or preferences), the point is this: whereas the econometric estimations might suggest that behaviour change in both categories would be best achieved through economic intervention, in fact there are more significant forces than price and income influencing expenditure in category B that – while hidden from the econometric analysis – may offer a more promising point of intervention.

The uncertainty is significant enough when the variables are treated at the aggregate level and the unobserved components represented in an abstract way. The uncertainty is compounded when consideration is given to how the social world is really constituted, including the structure of the unobserved components around the actions of individuals or households.

Aggregate Causality

In the earlier review of rational choice theory, one of the key criticisms discussed was the commitment to methodological individualism, where phenomona at the level of society must be explained in terms of the actions of individuals.

In ELESA, the data used for expenditure and income are aggregate data for all UK households, that is, for each quarter a single figure is given for all UK household income combined, another figure is given for all expenditure on gas, another for all expenditure on vehicle fuels and lubricants (and so on for all expenditure categories). Similarly, for price in each category, a single figure is given to reflect an implicit weighted average of all the price changes of different goods and services in that category. Since the variables are represented through aggregate data, the estimated relationships between those variables are also descriptions at the aggregate level.

If policy recommendations are to be drawn from ELESA about the effectiveness of economic interventions, then a commitment must be made regarding the causal efficacy of these aggregate relationships. That is, it must be believed that changes in one aggregate variable such as 'electricity price' will have an impact (based on the estimated relationship) on another aggregate variable, in this case 'electricity expenditure'.

Following the instrumentalist approach exemplified by Friedman, this commitment alone is sufficient so long as the aggregate relationships support successful prediction and intervention. No further commitment is required that might attempt to describe how the social world is constituted so as to bring about these changes. As Blaug (1992, p.99) argued though, the weakness of an instrumentalist approach is that – when it fails – there is little in the way of substantive theory to be adjusted and improved.

In uncovering the net effects of unobserved components driving UK household expenditure, ELESA offers an important piece of econometric analysis. However, the example given earlier showed how an identical net contribution from unobserved components for two categories could mask very different sets of underlying structures. Thus, ELESA also serves as an exemplar of why an instrumentalist philosophy is simply not up to the task of supporting a proper analysis of the social world. To understand the impact of economic intervention, it is necessary to give consideration to how the social world is really constituted.

Micro-level Causal Pathways

If some attempt must be made to give micro-foundations to the causal pathways between aggregate variables, a reluctant (and obviously false) individualist response might be to treat the aggregate effects of income, price and unobserved components as reflective of exactly the same effects holding at the level of the individual, that is, all individuals sharing identical incomes, tastes and preferences and exhibiting identical expenditure for each category of goods and services. In a case such as this, describing the causal pathway between the aggregate variables is simple: a change in one aggregate variable represents a set of (identical) micro-level changes, with (identical) micro-level effects that are in turn aggregated up to show a change in a second aggregate variable.

More reasonably, an individualist approach might acknowledge the obvious differences (in incomes, tastes and preferences, habits and routines and so on) between individuals, and yet still claim that a causal pathway can be traced through changes in distributions at the micro level of individuals, without recourse to intermediate social structures.

Criticisms of rational choice theory discussed earlier pointed to the role of emotions, habits, social structures and so on in influencing human behaviour. By now it should be obvious that if the unobserved components include 'all other factors' besides income and price, then even a very rough sketch of its contents must include factors internal to the individual (emotions, habits and so on) and factors external to the individual (social institutions, demographics, technical change). While the net effect of these factors has been estimated by ELESA, it would be a mistake to insist that these effects are best understood as the aggregation of effects at the micro level of the individual only.

Scientific Realism

For a proper understanding of the factors influencing consumer behaviour, a philosophical orientation is required that pays proper attention to how the social world is really constituted, that is, to the place of 'socially constructed and socially situated individual[s] within a set of local social relationships, institutions, norms, and rules' (Little, 2009).

This commitment to ontology forms the basis of scientific realism, a perspective introduced to the sciences by Bhaskar (1975 [2008]). More recently, the realist perspective has been brought to bear on economics (Lawson, 1997, 2003). Lawson's (1997, pp. 16–20) central criticism of mainstream economics – including econometrics – is its commitment to a mathematico-deductive style of reasoning. This presupposes closed systems that give rise to empirical event regularities (whenever x then y), much like the controlled experiments of the natural sciences. In econometrics, these event regularities are then the subject of statistical analysis.

By contrast, the realist perspective sees 'a world composed in part of complex things . . . which, by virtue of their *structures*, possess certain *powers* – potentials, capacities, or abilities to act in certain ways' (Lawson, 1997, p. 21, emphasis in original). While the components of the real world have the potential to act in certain ways, that is not to say they will.

According to the realist perspective, the world is stratified into three domains: the real, the actual and the empirical. Thus, whether or not a (real) power is exercised will determine the (actual) flow of events, and only some of those events will be subject to our (empirical) experience. It is this final domain of empirical facts that econometrics sets out to analyse. Yet, in the social world much more is actually going on besides. Furthermore, behind these actual events lie the real structures, powers, mechanisms that are the target of a realist scientific understanding.

While Lawson is sceptical of the possibility of rescuing econometric methods as part of a realist approach, others have offered a more conciliatory analysis. Downward and Mearman (2002) argue that although the social world is indeed open, individuals nevertheless seek security through alignment with social structures and institutions. That is, while individuals may have the power to act in unpredictable ways, they often revert to more or less predictable patterns of behaviour influence by habits, social norms and so on. Accordingly, 'regression-based analysis can contribute to the specification of "demi-regularities" from which causal analysis can proceed, drawing upon more historical and qualitatively based analysis' (Downward and Mearman, 2002, p. 407).

Any apparent regularity in the effects of income, price and unobserved components on expenditure, as estimated in ELESA, might be viewed in this way as a reasonable starting point for a more in-depth analysis of consumer behaviour. Such regularities are likely to be socio-historically limited and thus a poor basis for the predictive analysis of future behaviour. Instead, in the realist perspective, it is the structures and mechanisms underlying this behaviour that matter. A more thorough examination of these structures and mechanisms can begin to reveal the causal pathways by which changes in income, price and other factors can bring about change as part of an open, non-deterministic understanding of the social world.

Implications for Further Enquiry

The further development of ELESA might proceed in a number of ways. Firstly, there may be support for deriving policy recommendations on the basis of ELESA as it stands, without any consideration given to the nature of the unobserved components. Even in the presence of social structures, an instrumentalist commitment to aggregate causal pathways might still be maintained, by insisting that – however the changes come about – as long as the aggregate relationships hold, then an understanding of the underlying structures is unnecessary for effective policy making. It has been argued here that the presence of underlying trends in all 16 categories analysed in ELESA is a clear reminder that there is much more to understand about consumer behaviour than the role of income and price, but instrumentalist policy makers may disagree.

Alternatively, it may be that some of the unobserved components can be identified and subject to quantitative analysis of their own, for example, demographic and technological change, perhaps by endogenizing these variables into the model. This approach might be considered an attempt to make 'accommodations and adjustments' to reflect the presence of disturbing causes around a central cause of rationality. However far this endogenization might extend, the underlying structure of the remaining unobserved components would continue to raise uncertainty regarding the unintended consequences of any intervention.

A related approach would be to supplement the quantitative analysis provided in ELESA with a qualitative analysis of the unobserved components. Jackson (2005) discusses a range of alternate approaches to modelling consumer behaviour from the psychological and sociological traditions. One possibility might be to employ one or more of these models as under-labourers to ELESA in the manner discussed. However, the methodological critique of rational choice theory that gives rise to so many of these alternative models would suggest this is unlikely to be a comfortable fit.

Another approach would be to invert this relationship and make econometrics the under-labourer. This might be the only way to give proper consideration to the multitude of factors influencing consumer behaviour. Economic factors need no longer be treated as seperate and superior to the various disturbing causes. Rather, economic motivations at the level of the socially constructed and socially situated individual could be treated alongside other factors such as emotions and habits, social norms, physical infrastructures and so on. Econometric methods could then be used as required to explore specific questions, employing a variety of datasets in the process.

CONCLUDING REMARKS

The concept of rationality has been central to economics since its inception. The first political economists to consider methodology explicitly described rationality as an abstraction of human behaviour, applicable only to those areas of human conduct where wealth is the principle object (Mill, 1844, pp. 139–40). Even in those cases, it was believed that the concept must be supplemented by a consideration for disturbing causes.

Later, through the mathematization and quantification of economics, rational choice theory would eventually be taken up as a basic principle of consumer behaviour. Friedman (1953) encouraged economists not to be concerned with the 'realism' of their assumptions, but to treat them as instrumentally useful abstractions that could offer predictive power.

However, instrumentalism and rational choice theory have both come under sustained criticism from social scientists and methodologists. In a recent econometric analysis, the notion that consumer behaviour can be sufficiently described through economic variables alone has been challenged. ELESA applies a STSM to estimate the contribution of income, price and unobserved components to changes in UK household expenditure. The results of ELESA have been published elsewhere (Chitnis and Hunt, 2009a, 2009b, 2010, 2011), and while these are an important contribution to the literature on consumer behaviour, the methodological considerations raised in this chapter demonstrate the importance of addressing the underlying structures and mechanisms that influence consumers.

Lawson's deep rooted criticism of econometrics is its adherence to a positivist-deductivist orientation (Lawson, 1997). Although a reorientation of the philosophy of economics might be considered a distraction from the immediate task of assessing the contribution of ELESA, the scientific realism that Lawson proposes – with its concern for underlying structures, powers and mechanisms – addresses many of the same concerns raised in trying to make sense of the unobserved components.

The wider critique of econometrics goes back at least to Keynes, who wrote:

> If we were dealing with . . . independent atomic factors and between them completely comprehensive, acting with fluctuating relative strength on material constant and homogeneous through time, we might be able to use the method of multiple correlation with some confidence for disentangling the laws of their action . . . . In fact we know that every one of these conditions is far from being satisfied by the economic material under investigation. (Keynes, 1971, p. 286)

In a more conciliatory interpretation, following Downward and Mearman (2002), it may be that econometric analysis can be employed in the search for demi-regularities, patterns of consumer behaviour that – although limited within a particular region and time – can offer a starting point for a more substantive and mixed method approach to exploring the real structures of the social world.

ACKNOWLEDGEMENTS

I would like to thank Dr Mona Chitnis and Professer Lester Hunt, University of Surrey, UK, who developed ELESA and provided assistance in interpreting the results and implications of their work. I would also like to thank Professer Hunt for his helpful comments on an earlier draft of this chapter.

NOTES

1. The ESRC-funded research group ran between 2006 and 2011 at the University of Surrey, UK and included economists, psychologists, sociologists and political scientists. RESOLVE aimed to provide evidence-based advice to policy makers seeking to influence consumers' energy and environmental behaviour.
2. The categories of goods and services examined in ELESA come from the United Nations Classification of Individual Consumption According to Purpose (COICOP). The first version of ELESA adopted the 12 high level COICOP categories, but these were further disaggregated in later work to give: food and non-alcoholic beverages; alcoholic beverages, tobacco and narcotics; clothing and footwear; electricity; gas; other fuels; other housing; furnishings, household equipment and routine maintenance of the house; health; vehicle fuels and lubricants; other transport; communication; recreation and culture; education; restaurants and hotels; miscellaneous goods and services.
3. The contribution of average annual temperature was also considered in the case of electricity, gas and other fuels.
4. This chapter will not consider specific policy measures, although it is worth noting that the impact of changes in household income could not be confined to any particular category.

REFERENCES

Bhaskar, R. (1975), *A Realist Theory of Science*, York: Leeds Books, reprinted in 2008, London: Verso.

Blaug, M. (1992), *The Methodology of Economics: Or How Economists Explain*, 2nd edn, Cambridge: Cambridge University Press.

Caldwell, B. (2003), *Beyond Positivism*, London: Routledge.

Chitnis, M. and L. Hunt (2009a), 'Modelling UK household expenditure: economic versus non-economic drivers', RESOLVE Working Paper Series 07–09, available at http://resolve.sustainablelifestyles.ac.uk/publications (accessed 2 July 2013).

Chitnis, M. and L. Hunt (2009b), 'What drives the change in UK household energy expenditure and associated CO2 emissions, economic or non-economic factors?', RESOLVE Working Paper Series 08–09, available at http://resolve.sustainablelifestyles.ac.uk/publications (accessed 2 July 2013).

Chitnis, M. and L. Hunt (2010), 'Contribution of economic versus non-economic drivers of UK household expenditure', RESOLVE Working Paper Series 03–10, available at http://resolve.sustainablelifestyles.ac.uk/publications (accessed 2 July 2013).

Chitnis, M. and L.C., Hunt (2011), 'Modelling UK household expenditure: economic versus noneconomic drivers', *Applied Economics Letters*, **18**, 753–67.
Downward, P. and A. Mearman (2002), 'Critical realism and econometrics: constructive dialogue with post Keynesian economics', *Metroeconomica*, **53**, 391–415.
Elster, J. (1989), *Nuts and Bolts for the Social Sciences*, Cambridge: Cambridge University Press.
Friedman, M. (1953), *Essays in Positive Economics*, Chicago, IL: University of Chicago Press.
Harvey, A.C. (1991), *Forecasting, Structural Time Series Models and the Kalman Filter*, Cambridge: Cambridge University Press.
Hausman, D.M. (2008). 'Philosophy of economics', *Stanford Encyclopedia of Philosophy*, available at http://plato.stanford.edu/entries/economics/ (accessed 2 July 2013).
Heath, J. (2011), 'Methodological individualism', *Stanford Encyclopedia of Philosophy*, available at http://plato.stanford.edu/entries/economics/ (accessed 2 July 2013).
Hutchison, T.W. (1938), *The Significance and Basic Postulates of Economic Theory*, London: Macmillan.
Jackson, T. (2005). 'Motivating sustainable consumption: a review of evidence on consumer behaviour and behavioural change', available at score-network.org/files/843_23.pdf (accessed 2 July 2013).
Keynes, J.M. (1971), *The Collected Writings of John Maynard Keynes: The General Theory and After: Part 2, Defence and Development*, London: Royal Economic Society (Great Britain), Macmillan.
Lawson, T. (1997), *Economics and Reality*, London: Routledge.
Lawson, T. (2003), *Reorienting Economics*, London: Routledge.
Little, D. (2009), 'Methodological localism', available at http://understandingsociety.blogspot.co.uk/2009/11/localism-and-assemblage-theory.html (accessed 2 July 2013).
Mill, J.S. (1844), *Essays on Some Unsettled Questions of Political Economy*, London: John W. Parker.
Mill, J.S. (1848), *Principles of Political Economy with Some of their Applications to Social Philosophy*, new edition introduced by Sir W.J. Ashley, London: John W. Parker.
Robbins, L. (1932). 'An essay on the nature and significance of economic science', available at http://www.scribd.com/doc/14242989/An-Essay-on-the-Nature-and-Signicance-of-Economic-Science-Lionel-Robbins (accessed 2 July 2013).
Samuelson, P. (1947), *Foundations of Economic Analysis*, Cambridge, MA: Harvard University Press.
Senior, N.W. (1836), *An Outline of the Science of Political Economy*, London: W. Clowes and Sons.
Simon, H.A. (1957), *Models of Man*, Oxford: Wiley.
Weber, M. (1956) *Wirtschaft und Gesellschaft. Grundrisse der verstehend en Soziologie*, 4th German edn, edited by J. Winckelmann (ed.), Tubingen: J.C.B. Mohn (P. Siebech), pp. 1–550, 559–822, reprinted in 1978, *Economy and Society*, San Francisco, CA: University of California Press.

4. Scenarios as tools for initiating behaviour change in food consumption

Andrea Farsang and Lucia Reisch

INTRODUCTION

Over the last few decades, considerable changes in food consumption have taken place. The food system is a complex socio-ecological system surrounded by unpredictable events and uncertainties, especially on long-term horizons such as 25–50 years ahead. In order to deal with these uncertainties, complexities and long-term challenges as well as to influence developments proactively, scenario planning is increasingly applied in both policy making and knowledge brokerage. Scenarios can reveal uncertainties, can help prepare for unexpected changes and highlight crucial decisions to be taken today. Scenarios can offer a clearer picture of the present and visions for the future and can help to identify key driving forces and barriers for changing behaviour and their trends as well as assess potential outcomes of different policy paths.

The intention of this chapter is to provide background information and a short summary on the use of scenarios in evidence-based policy making in general as well as in the context of changing behaviour towards sustainable food consumption. Moreover, it compiles and briefly reviews recent examples of scenario building approaches in the sustainable food domain. The potential roles and applications of different types of scenarios such as visioning exercises, backcasting or quantitative models and their benefit for policy making are discussed. The chapter is built on a discussion paper that was prepared for the 'Policy Meets Research' Workshop on sustainable food consumption within the European Union (EU)FP 7 CORPUS project that-took place in Vienna in 2011 and 2012 (http://www.scpknowledge.eu).

DRIVING FORCES AND FUTURE TRENDS IN FOOD CONSUMPTION

There have been considerable changes in food consumption – particularly in relation to eating habits, dietary changes, nature and quantity of food – over the last few decades mainly due to an increase in agricultural productivity, greater diversity and less seasonal dependency. Through falling food prices and rising incomes, food has become more affordable in many parts of the world. Yet, while there have been significant improvements in decreasing under-nourishment globally (Alexandratos, 2006), there are still countries where calorie intake has declined. Due to the food price spike in 2008–09 the share of people suffering from hunger has grown. According to the Food and Agricultural Organization (FAO) (2009), food production has to increase by up to 70 per cent in order to feed the growing population – projected to be around nine billion people by 2050 – with the biggest increase in low-income countries (Africa is projected to double its population by 2050). It is a huge challenge for agriculture to meet this growing demand for food in a sustainable way. Other important future challenges – discussed at length elsewhere, for example, Reisch et al. (2013) – are nutrition transition, diet-related health problems, food related uncertainty and distrust (for example, pesticides, hormones, antibiotics, additives and genetically modified organisms (GMOs), rising level of urbanization, environmental impacts and (future) governance of the food system.

Currently, half of the world population lives in urban areas (UNFPA, 2010) and urban populations are expected to reach 6.3 billion in 2050 (UNPD, 2009) with most significant increases in developing countries. Rapid urbanization has severe effects on food consumption patterns. In a comparison of rural and urban diets, the urban population shows a higher intake of calories, fat, animal protein, sugar and prepared food (Popkin and Gordon-Larsen, 2004), which is mainly due to an increase in convenience food and out-of-home consumption (The Government Office for Science, 2011a). This is exacerbated by excessive food advertising by the food industry (The Government Office for Science, 2011b) and by rising income levels resulting in decreased relative consumer expenditure on food in many European countries (EEA, 2005). At the same time, after the food price crisis in 2007–08, prices started to rise again from 2009 reaching a record high in 2011 according to the FAO's world food price index, driving around 44 million more people into poverty in low- and middle-income countries (World Bank, 2011). Due to the financial crisis, in many European countries the use of public food banks increased considerably (European countries in 2011 faced a 20 per cent growth in use of food banks, cited in UNECE, 2012) while half of edible and healthy food is wasted in households and supermarkets (in the EU-27 89 tonnes a year, 179 kg per person yearly). A related challenge is nutrition transition that reflects changes in diet – characterized by higher intake of meat, sugar, saturated fat, salt and low consumption of vegetables and fruits – lower level of physical activities and related health problems. Nutrition transition describes a shift from undernourishment to nutrition-related non-communicable diseases (NR-NCDs) such as obesity or cardiovascular diseases, diabetes and cancer (Popkin, 2004). The trend towards higher consumption of fish and meat is expected to continue; for meat estimates are a rise from 37 kg to 52 kg per person per year by 2050 (FAO, 2009).

Along with obesity, malnutrition – referring to both under- and over-nourishment – is a growing concern in the highest socioeconomic status categories in low-income countries and in socially and economically disadvantaged groups of society (SES) in industrialized countries (Devaux and Sassi, 2011; Popkin, 2002) especially of the urban population. Mainly affected groups are the elderly, children, poor and sick people. Malnutrition concerns the amount and type of food intake and describes a situation of imbalance between supply and demand of nutrients and energy needed to maintain different functions of the human body and to ensure normal growth (WHO, 1996). Malnutrition occurs predominantly in so-called 'food deserts': areas of relative exclusion where people experience physical and economic barriers to accessing healthy foods (Reisig and Hobbiss, 2000).

Another challenge to tackle is that consumers are increasingly uncertain and distrustful of food suppliers. This is mainly due to recent and reoccurring food scares in Europe and the growing distance between consumers and producers. Substantial changes are observable in future governance of the food system at both national and international levels. Due to the concentration of food processing, distribution and retailing price competition has become enormous (see Reisch et al., 2010). A limited number of transnational corporations control food retail sales. According to the report *Foresight. The Future of Food and Farming* (The Government Office for Science, 2011a), this trend may slow down when new companies from emerging economies enter the markets. Hence, it is of major importance how production subsidies, trade restrictions and other market interventions will develop in the future.

Last but not least, emerging environmental problems such as climate change, land degradation, loss of biodiversity, accelerating global energy and water demand are closely linked to food production, consumption and distribution. According to the European Environment Agency (EEA, 2011, p. 2), 'global food, energy and water systems appear to be more vulnerable and fragile than was thought a few years ago, due to increased demand for food, and a decreased and unstable supply'. Major contributors to environmental problems are water used for production, use of fertilizers (with impacts on groundwater, soil and air), intensification of production (and hence loss of agro-biodiversity), increasing food miles and food waste. To date, it is an unsolved puzzle how markets will be able to meet the growing food demand without unduly compromising environmental quality. Alternative forms of regional, local and urban food production, distribution and consumption, changes in diets towards less meat and more vegetables as well as a move towards organic farming are strategies worth exploring.

Preparing scenario exercises, these trends and expected challenges in the field of sustainable food consumption primarily revolve around questions relating to:

- Drivers: how will potential drivers develop in the coming decades? How will they interact with each other? How will they affect the (un) sustainability of the food system?
- Goal conflicts: what are the trade-offs between the different goals and aims – food security, environmental concerns and so on – in envisioning sustainable food futures?
- Systems view: what changes would occur in the food system if one of the variables is changed (for example, more organic food)? How would the food system respond to drastic changes or crises in any of the above areas (food scares, energy crisis and so on)?
- Governance: how do we envision future food governance? How does it affect volatilty, global and local developments?
- Glocalization: how do global developments and local preferences interact?

SCENARIO TYPES AND EXAMPLES FOR BEHAVIOUR CHANGE

The aims of this chapter are twofold: firstly, to explain the use, merits and limits of scenario building as a policy tool in general; secondly, to give an overview of recent scenario studies on sustainable futures in the food sector.

Scenario Building as a Policy Tool: Scope and Types

Contemporary social and environmental problems call for systemic, structural changes towards global sustainability in different sectors, particularly in energy, mobility and the food sector (Elzen et al., 2004; Grin et al., 2010; Van den Bergh and Bruinsma, 2008). As problems in these domains are highly complex and uncertain, we need complex and long-term processes of transition in order to sustain the way we fulfil societal needs (Raskin et al., 2002). Such transitions require changes at different levels and also need to incorporate the involvement of multiple stakeholders. Transitions towards sustainability refer to a 'radical transformation towards a sustainable society as a response to a number of persistent problems confronting contemporary modern societies' (Grin et al., 2010, p. 1).

In transition management research, the multi-level perspective (MLP), originally developed by Rip and Kemp (1998) and further elaborated by Geels, Grin, Kemp, Rotmans and others (Geels, 2005; Grin et al., 2010; Kemp et al., 2011; Rotmans et al., 2001; Smith et al., 2005; Verbong and Geels, 2007), analyses transition dynamics by distinguishing three levels within the societal system: the 'landscape', representing macro-level trends, barriers and drivers, the 'regime', which represents dominant institutions and technologies and 'niches', in which radical innovations emerge or have the potential to emerge (de Haan and Rotmans, 2011; Geels and Schot, 2007; Grin et al., 2010), providing seeds for change.

Social practice theory offers a promising complementary approach when studying and explaining individual behaviour and agency and its influencing factors, both external and internal such as social norms, infrastructure and systems of provision, institutions and practices (Shove, 2003; Warde, 2005).

Scenario development evolved as a strategy tool originally applied in military planning, developed by Herman Kahn and his colleagues at the RAND Corporation (Van der Heijden et al., 2002). The first generation of scenarios, developed in the 1950s, were mainly statistical predictions (Slaughter, 1996). In the 1970s, a second wave of scenarios began to receive attention by corporate planners at Shell and General Electric who noted the deficiencies of forecasts as these often proved to be incorrect especially in the long run. Another well-known example of scenarios in the second wave is *The Limits to Growth* report to the Club of Rome in 1972 (Meadows et al., 1972). In the 1980s, a new, third generation of scenarios was developed in the context of sustainability challenges. These recognized the need for societal and structural changes, a transition in order to find the way to a more sustainable future as well as new methods for explorations of the future (Sondeijker, 2009).

There are various definitions of scenarios and scenario development (for example, Dreborg, 2004; Gallopin et al., 1997; Tapio and Hietanen, 2002; Van Notten et al., 2003). Beyond all differences, there is a general agreement that scenarios are not predictions or projections (Rotmans et

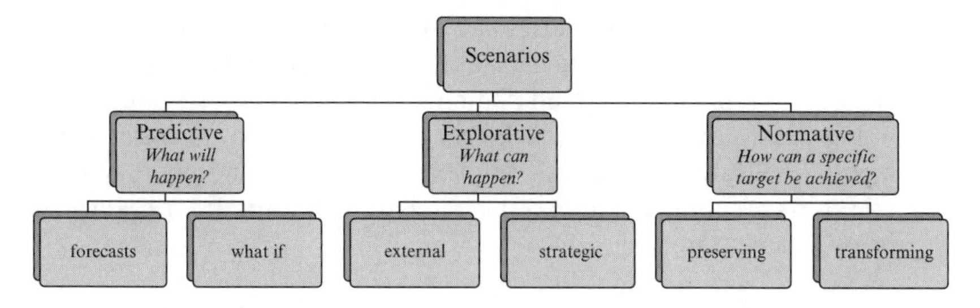

Source: Following Börjeson et al. (2005).

Figure 4.1 Typologies of scenarios

al., 2000; Van Notten et al., 2003). Scenarios rather describe alternative images of the future with the assumption that future developments are unpredictable and stress the need to take uncertainty into account in decision making. Van Notten (2005, p. 2) proposes a definition that covers most aspects of the various approaches around the definition of scenarios:

> Scenarios are consistent and coherent descriptions of alternative hypothetical futures that reflect different perspectives on past, present, and future developments, which can serve as a basis for action.

As regards the different characteristics of scenarios, several typologies have been proposed (for example, Börjeson et al., 2005; Dreborg, 2004; Slaughter, 1988; Tapio and Hietanen, 2002, Van Notten et al., 2003). A particularly useful typology has been developed by Börjeson and colleagues (2005), proposing three general types of scenario approaches coined 'predictive', 'explorative' and 'normative' (Figure 4.1):

1. Projections as predictive tools answer the 'what will happen?' question (forecasts and 'what-if' scenarios predominantly quantitative).
2. Exploratory scenarios answer the 'what can happen?' question (external or strategic, typically qualitative, aim to explore plausible futures and develop a set of scenarios on a long time horizon).
3. Normative scenarios answer the question: 'how can a specific target be reached?' There are two types: (in 'preserving' scenarios, the targets can be reached without transformation; they mainly work with optimizing modelling or in a qualitative way. 'Transforming' scenarios have to be used if structural changes in the system are needed, such as 'backcasting'; these typically result in a number of images or visions of the future illustrating how specific outcomes or a certain target can be reached. Quist and Vergragt (2006, p. 1028) define backcasting as

> 'first creating a desirable (sustainable) future vision or normative scenarios, followed by looking back at how this desirable future could be achieved, before defining and planning follow-up activities and developing strategies leading towards that desirable future'. Backcasting is highly applicable in cases of complex problems, when major, systematic shifts are needed and current trends and lock-ins are necessary to be overcome (Dreborg, 1996).

Contemporary scenario development and techniques are used in a wide range of contexts and in different organizational settings from corporate planning to public policy assessment. As scenario development is applied to respond to different challenges and developments, the methods and techniques used vary according to the nature of change addressed (for a detailed review see Bishop et al., 2007).

Scenarios are expected to serve a wide range of functions in both policy making and knowledge brokerage. In particular, according to Börjeson et al., (2005), Swart et al. (2004), Van der Heijden (1996) and Van Notten et al. (2003), they have the potential to:

- Provide a more accurate picture of the present and identify uncertainties. Through analysing different policy strategies via scenarios, decision makers are able to assess the differing outcomes and the key driving forces that influence those outcomes.
- Help discover existing problems or emerging, uncertain aspects of the future, expand creativity, identify opportunities and threats and explore possible ways to respond to these.
- Improve policy strategies; different scenarios can describe the probable results of different actions and policies and contribute to the debate on different policy options.
- Enhance consensus building and increase the level of social learning, which is particularly the case for participatory scenario-planning processes through public and stakeholder participation in the formulation and evaluation of scenarios.
- Visualize images of a desirable future pointing to different lifestyles, values, viewpoints and perspectives, the way we consume, cultural shifts and options for behaviour change.
- Incorporate the normative aspects of sustainability including values, behaviours, culture and institutions.

As regards methodological approaches of scenario development, the literature distinguishes between desk research, model-based and participatory approaches (Van Notten, 2005). Participatory methods mainly work with stakeholder workshops, focus groups, citizens' juries and envisioning workshops to reach specific aims. Therefore, through public, expert and stakeholder participation, scenarios can broaden the perspective, information can be collected from a wide range of disciplines and can be both challenged and integrated (Rotmans, 1998), linking scientific knowledge and political decision-making (Millenium Ecosystem Assessment, 2005). Moreover – and of increasing political importance – public and/or expert involvement in scenario formulation and evaluation can increase legitimacy and acceptance of visions and political decisions among participants and beyond (Quist, 2009).

Scenarios are increasingly used to respond to and influence development (Sondeijker, 2009), to discover uncertainties, to prepare for unexpected changes and to 'highlight crucial decisions that should be taken today' (Mutombo and Bauler, 2009, p. 1). In order to contribute to these goals, using scenario techniques needs careful planning. There are a range of basic questions that have to be addressed before entering a scenario exercise:

- Target group and goals: for whom should scenarios be designed (decision makers, researchers, community)? What types of scenarios are the most 'user-friendly' among different groups of 'scenario users'? What is the use and purpose of different future studies for different groups of users?
- Scope: what should be the focus and depth of scenario building: in-depth focus on the food system itself or rather focusing more on the context of a sustainable food system?
- Methods: what should be the 'ratio' between model-based and more qualitative scenario building in evidence-based policy making?
- Participation: on what level and in what form is participation most effective in scenario building? When do we see participatory scenario building as successful?

SCENARIOS FOR SUSTAINABLE FOOD FUTURES: EXAMPLES AND STUDIES

The past decade has seen quite a number of studies investigating the future of the global food system (Reilly and Willenbockel, 2010). These studies utilize mainly long-term model-based simulations or participatory backcasting. This chapter provides a brief overview of a few recent national and international scenario excercises in the sustainable food domain (short descriptions are provided for projects applying backcasting and participatory envisioning), starting from micro regions and national to European and global scenario studies and examples. Table 4.1 provides a condensed overview.

Gothenburg 2050 (2002)

The Gothenburg 2050 project (see http://www.goteborg2050.se) developed a three-step scenario building method through the use of backcasting by an analysis of the current state of society and trends, developing the criteria for a sustainable city in a sustainable society and, finally, envisioning images of the future through participatory workshops. At the last stage, visions are compared to the current situation and trends and are used to set up 'guidelines' for change. The backcasting method has five features for sustainable food scenarios:

- sustainable and locally produced food
- a diet with a higher proportion of vegetables
- shorter distance and closer relations between producers and consumers
- food trade placed in local squares
- conscious and energy-efficient consumption.

The Gothenburg project was quite successful in engaging stakeholders and raising awareness of the different topics listed above (University of Pennsylvania, case studies, http://upenn-envs667660.webs.com/Case%20studies/Goteborg.pdf). However, the project has not yet managed to secure favourable policy decisions. Not surprisingly, it seems to be a difficult trade-off between allowing participants maximum freedom in developing future scenarios and translating the outcomes of backcasting directly into policy decisions. One lesson to be obtained from this is that in scenario building, it is important to distinguish between short-term actions and longer-term objectives – the latter should inform short-term decisions but should not require a longer time to implement.

Strategies and Scenarios for Managing the Transition to Sustainable Food Consumption: Elements from the CONSENTSUS Project (2008)

The Belgian CONSENTSUS project (Boulanger, 2008) aimed to test and assess the potential of a transition management approach to sustainable consumption. The project included three participatively constructed backcast scenarios as the first step: eco-efficiency, de-commoditization and sufficiency scenarios. Since food consumption is a mixture of the three

Table 4.1 Examples of scenarios

Scenario project	Issues covered	Methodology	Region covered
Gothenburg 2050 (2002)	food, transportation and energy use related to food	normative, participatory backcasting	small region (Gothenburg)
Risku-Norja (2011)	environmental impacts of primary consumption	explorative, model based	small region (Finland)
Vinnari (2010)	meat consumption	normative, backcasting	country (Finland)
CONSENTSUS (2008)	food consumption	normative, participatory backcasting	country (Belgium)
Belgian Federal Report on Sustainable Development (1999, 2003, 2005)	sustainable development in general, food	explorative, normative	country (Belgium)
Livewell (2011)	greenhouse gas emissions (GHGE)	normative, backcasting	country (UK)
Chatham House (2008)	global food supply	explorative, model based	UK and EU
CORPUS (2010–12)	food, mobility, housing	normative, backcasting	EU, global
CONSENSUS (2007–13)	food, energy, mobility and water	normative, backcasting	country (Ireland)
Getting into the Right Lane for EU 2050 (2009)	food and biodiversity, land resources	normative, backcasting	EU
SPREAD 2050 (2011–12)	food, mobility, housing, health and society	normative, backcasting	EU, global

Table 4.1 (continued)

Scenario project	Issues covered	Methodology	Region covered
CRISP (2011–14)	food, mobility, energy/housing	normative, backcasting	EU, global
Stehfest et al. (2009)	impact of livestock sector on GHGE, land use and carbon cycle	explorative, model based	global
Jensen and Smed (2007)	impact of financial tools on nutrition	explorative, model based	global
The fourth Global Environment Outlook entitled Environment for Development (GEO-4) (2007)	sustainable development in general	explorative	global
OECD Environmental Outlook (2008)	climate change, biodiversity loss, water scarcity and health	predictive, explorative	global
Agrimonde (2009)	food consumption, land use, resource usage	predictive, normative	global
Millenium Ecosystem Assessment (2005)	ecosystem and human well-being	explorative, model based	global
Foresight. Tackling Obesities (2007)	food, obesity	explorative, model based	global
Foresight. The Future of Food and Farming (2011)	food prices, hunger	explorative, model based	global

paradigms above, the second step was to construct an integrated scenario of food consumption.

Thinking about the Future of Food/The Chatham House Food Supply Scenarios: Chatham House Food Supply Project (2008)

The Chatham House Food Supply Project (2008) developed four global food supply scenarios to examine their impact on the UK and the (EU). Based on previous research, four dimensions were taken into consideration while constructing the scenarios: (1) the changing oil price, (2) the growth of global demand on food; (3) the supply capacity; and (4) global and economic answers to the three topics. Storylines were developed in a participatory way with research teams and experts in the given topic. The starting points of the scenarios were the following:

- Just a blip: what if the current high price of food is to end in a short time and food will be cheap again?
- Food inflation: what if food prices remain high for several years?
- Into a new era: what if the current food system has reached its limits and has to change?
- Food in crisis: what if a major food crisis develops due to diseases and water shortage linked with extreme energy prices due to the geopolitical situation?

CORPUS: Enhancing the Connectivity between Research and Policy-making in Sustainable Consumption (2010–13), 3rd Discussion Paper on Sustainable Food Scenarios (Reisch et al., 2011a, 2011b)

This aimed to develop novel approaches to knowledge brokering between policy making and research in three areas of sustainable consumption: food, mobility and housing. It was facilitated by a web platform and interaction exercises, including collaborative scenario building and backcasting workshops with the aim to identify alternative future pathways and help policy makers discover key driving forces, uncertainties and diverging views to enhance convergence and mutual understanding. Reflecting on the key challenges of sustainable food consumption and behaviour change, the following issues were chosen to be given special attention: a diet with a higher proportion of vegetables; alternative food networks (AFNs) and short food supply chains (SFSCs); public procurement; and conscious and energy-efficient food consumption. The sub-scenarios developed by small groups of researchers and policy makers and described below represent a desirable food vision as a:

- public procurement sub-scenario: use of public canteens to create a window of opportunity for change both in the food chain and on the demand side
- sustainable and locally produced food sub-scenario: reconnecting consumers and producers, promotion of local, traditional farms, urban agriculture, community gardens and so on
- healthy and sustainable diet sub-scenario: combating obesity and unhealthy diets and promoting more healthy and sustainable ways of eating
- high-tech food sub-scenario: exploring the way for a controlled and safe development of high technology in food agriculture and the food industry while meeting sustainability requirements
- quality/enjoyable food sub-scenario: promoting seasonal, local and high-quality food with lifestyle changes (Slow-Food approach)
- new social norms sub-scenario: use of both campaigning and banning unsustainable products in order to induce long-term changes in food consumption practices.

The majority of scenarios developed in the workshop approach the domain in a holistic or global perspective by either focusing on the whole food system or dealing with it from a multidisciplinary perspective. In almost all visions, we can observe a challenging tension, namely, between global and local solutions. Other important aspects and issues raised in the development of visions were to reflect the real costs of food and quality instead, target a radical change in communication towards transparency and clear labels and a more cultural approach in order to change culture-specific consumption patterns (Reisch et al., 2011b).

CONSENSUS: A Cross-border Household Analysis of CONSumption, ENvironment and SUStainability in Ireland (2007–13)

The four year collaborative research project CONSENSUS deals with the four key consumption domains having the largest impact on the environemnt: energy, mobility, water and food. The project focuses on Ireland but provides results applicable on the European level, dealing with the measurement of consumption, sustainable behaviours, identification of the links between consumption, health and well-being as well as institutional practices and policy recommendations. Within the project, three scenarios were developed through visioning workshops in the food domain with three distinctive levels of focus: people, technology and organization (Pape, 2011):

'Come together to eat'/People, 'Smart kitchens'/Technology and 'Educate and incentivize'.

SPREAD: Sustainable Lifestyles 2050 (2011–12)

The SPREAD Sustainable Lifestyles 2050 European social platform project aims to develop a vision for sustainable lifestyles in 2050 by bringing together in dialogue a wide range of societal stakeholders and, as a result, setting up a roadmap for policy makers and generating innovative ideas for making a transition towards sustainable lifestyles.

In the frame of the project, four scenarios have been developed with the use of backcasting, along two uncertainties or critical variables: technology (either pandemic or endemic) and society's governing principle (human-centric or meritocratic). After defining four scenario landscapes (see the short scenario descriptions below), pathways have been explored through backcasting workshops, then quantified and qualified and finally visualized.

- Singular Super Champions: is the business opportunity of the century. Price and a health-efficient diet and large-scale organic production dominate.
- Governing the Commons: new forms of collaboration develop and digital reality redefines our lifestyles, norms and relationships. Food production and distribution are managed by global food systems. Vegetable choices and synthetic meats form differentiations in diets.
- Local Loops: Energy and resource use is managed through local and regional cycles. Technology is adapted to local circumstances and needs, and a new kind of professionalism arises. Food production and distribution are marked by locality, minimized transportation and neighbourhood canteens.
- Empathetic Communities: The community and neighbourhood have an important role in daily life and solutions are developed at the local level. Growing food in urban farming circles meets local food demand. Food transportation needs are very low. High importance is also placed on food quality and distribution.

CRISP: CReating Innovative Sustainability Pathways (2011–14)

The project as a whole aims to outline bottom-up approaches that address the question of how to overcome the gap between awareness of the issues at stake and the concrete engagement in sustainability-driven action, as individuals and as a society. The project develops visions of a low-carbon, sustainable life in 2030 and provides insights on which transition pathways might best faclilitate the shift towards a post-carbon society. It also reviews 30 initiatives by looking at the transferability and up-scaling possibilities of the case studies.

The project involves a wide range of stakeholders – with a special focus on 17–19-year pupils – in scenario and backcasting workshops in six European countries. The three distinctive scenarios constructed by the pupils were based around the following: focus on local communities; I-Tech dominated scenario; and 'one ethical world'.

Getting into the Right Lane for EU 2050 (2009)

Using a backcasting methodology including land resources, food and biodiversity, energy and climate change as well as transport and mobility, the report investigates the challenges for EU policy up until 2050 (Sedlacko and Gjoksi, 2010). The vision for land resources, food and biodiversity is to be able to produce food for nine billion people, while minimizing impacts on ecosystems. Four scenarios of possible future global roles for the EU in relation to these challenges are developed:

Europe as a superpower (strong focus on sustainable development); Europe is globalized, with strong international cooperations both in political and economic areas; Europe as a mercantilist, where international cooperation is limited; and Irrelevant Europe, isolated and out of international cooperations due to co-existing national and regional levels.

Agrimonde Scenarios and Challenges for Feeding the World in 2050, INRA-CIRAD 2009 (2009)

The aim of the Agrimonde project (INRA-CIRAD, 2009) was to set up scenarios that looked at feeding the world in 2050. For this purpose, stakeholders developed two scenarios. The first scenario, Agrimonde GO, was a trend-based scenario taking economic and technological development as a priority in order to forecast food consumption, land use and resource usage within current standards and policy framework. The second scenario, Agrimonde 1, was a normative scenario based on the vision of a sustainable food and agricultural system in 2050. Both scenarios were based on the same population growth forecast, which enabled a comparison of policy decisions needed for a sustainable future. A major outcome of the study is that both Agrimonde GO and Agrimonde 1 scenarios would provide enough food but Agrimonde GO would have significant consequences for the environment.

The Millennium Ecosystem Assessment (2005)

The project developed four plausible scenarios that considered increasing globalization or increasing regionalization as two possible paths of world development and then adopted either a reactive or a proactive approach to their ecosystem management:

- The Global Orchestration scenario, which depicts a worldwide connected society in which global markets are well developed and supranational institutions are well placed to deal with global environmental problems. The scenario highlights the risks from ecological surprises.
- Order from Strength scenario represents a regionalized and fragmented world concerned with security and protection, paying little attention to the common goods, and with an individualistic attitude towards ecosystem management. People in this scenario are also confident about technological development to solve environmental problems.
- The Adapting Mosaic scenario designs a fragmented world resulting from discredited global institutions. Communities slowly realize that local environmental issues cannot be treated locally due to the global nature of the environment and start to develop real, working networks among communities and nations for better management.
- The Technogarden scenario builds on a globally connected, technology-oriented world.

The findings of the Millennium Ecosystem Assessment study indicate that the degradation of the ecosystem in order to meet increasing demands can only be reversed if substantial changes in policy and practice are made urgently.

CONCLUSION AND IMPLICATIONS

The food system is a complex socio-ecological system relating to social, economic, health and environmental issues facing a high degree of uncertainty, especially in the long-term future such as 20–50 years. In order to deal with uncertainties, complexity and long-term challenges, scenario planning is increasingly applied (Raskin et al., 2002). Effective policies need to be established with support from practices discussed that enable a viable and effective route to transition. Scenario planning helps to indentify alternative future pathways and can help policy makers discover the key driving forces and uncertainties. When this information is linked with scientific knowledge and stakeholder participation, to legitimize and encourage acceptance, the decision-making process is more robust.

Participation offers possibilities to include values, behaviours, cultures and institutions and the normative aspects of sustainability.

Based on the review of scenario examples, we can conclude that scenarios, by definition, are limited to including a few selected parameters rather than the whole food domain, for example: GHGE, land use or hunger reduction. Territorial scales range from local to global including micro regions and countries. Model-based scenarios provide quantified data and can be used anywhere if given appropriate data, while backcasting scenarios focus more on supporting policy making. With respect to participatory backcasting scenarios, it is important to emphasize that limiting the area of discussions also limits the scope of vision, while too much liberalization of the discussion makes policy making more difficult.

The scenario study and project examples show the importance of involving a wide range of stakeholders in the development and backcasting of sustainable future visions in a participatory way for transition processes. It offers the possibility to incorporate the views and opinions of various stakeholders in strategic planning, policy making and the construction of action plans (for example, in the cases of CORPUS, CRISP, SPREAD and CONSENSUS) as well as to increase the legitimacy of visions and policy actions. The involvement of consumer-citizens (such as in the cases of CONSENSUS and CRISP) helps to test the acceptability of practices, policies and action plans and explore what kind of actions and interventions for behaviour change work in a given cultural context, how different actors, especially consumers and younger generations (for example, the CRISP project) conceptualize the good life and sustainable lifestyles. Understanding their different viewpoints, motivations and fears, obstacles and opportunities might help to overcome implementation challenges and increase the legitimacy of actions to be taken today. It can increase the effectiveness of scenarios aiming to support policy making if the system is analysed from an actor-based view and the constructed long-term pathways are connected to decisions to be taken up by the different actors in the near term. Complementary to the study of systematic transitions, social practice theory offers the possibility to explore the complex interactions towards lifestyle change. In participatory scenario and visioning workshops, the desirable futures visualized provided insights about values, lifestyles, driving forces and perspectives as well as possible ways for behaviour change and the decisions that have to be taken today.

ACKNOWLEDGEMENT

The chapter is built on the discussion paper 'Scenario developed for sustainable food consumption', prepared for one of the three 'Policy Meets Research' workshops on sustainable food consumption within the CORPUS Consortium (http://www.scp-knowledgeu). This research was funded by the European Commission, fp 7, project No. 244103 and carried out on behalf of the CORPUS Consortium.

REFERENCES

Alexandratos, N. (2006), *World Agriculture: Towards 2030/50, Interim Report. A FAO Perspective*, Rome: FAO.

Bishop, P., A. Hines and T. Collins (2007), 'The current state of scenario development: an overview of techniques', *Foresight*, **9** (1), 5–25.

Börjeson, L., M. Höjer, K.H. Dreborg, T. Ekvall and G. Finnveden (2005), *Towards a User's Guide to Scenarios – a Report on Scenario Types and Scenario Techniques*, Stockholm: Royal Institute of Technology.

Boulanger, P.M. (2008), 'Strategies and scenarios for managing transition to sustainable food consumption: elements from the CONSENTSUS project', paper presented at the International Conference on Sustainable Consumption and Alternative Agri-food Systems, Arlon, 28–30 May.

De Haan, J. and J. Rotmans (2011), 'Patterns in transitions: understanding complex chains of change', *Technological Forecasting and Social Change*, **78** (1), 90–102.

Devaux, M. and F. Sassi (2011), 'Social inequalities in obesity and overweight in 11 OECD countries', *European Journal of Public Health*, first published online 6 June 2011, doi: 10.1093/eurpub/ckr058.

Dreborg, K.H. (2004), *Scenarios and Structural Uncertainty*, Stockholm: Department of Infrastructure, Royal Institute of Technology.

EEA (European Environment Agency) (2005), *Household Consumption and the Environment*, Report No. 11/2005, Copenhagen: EEA.

EEA (European Environment Agency) (2011), *Food for Thought – Sharing Information on Food Production*', available at http://www.eea.europa.eu/articles/food-for-thought (accessed 8 January 2011).

Elzen, B., F.W. Geels and K. Green (eds) (2004), *System Innovation and the Transition to Sustainability: Theory, Evidence and Policy*, Cheltenham, UK and Northampton, MA, USA: Edward Elgar.

FAO (Food and Agriculture Organisation) (2009), *Proceedings of the Expert Meeting on How to Feed the World in 2050: 24–26 June 2009*, Rome: FAO.

Gallopin, G., A. Hammond, P. Raskin and R. Swart (1997), 'Branch points: global scenarios and human choice', Resource paper for the Global Scenario Group, Stockholm Environment Institute, Stockholm.

Geels, F. (2005), 'The dynamics of transitions in socio-technical systems: a multi-level analysis of the transition pathway from horse-drawn carriages to automobiles (1860–1930)', *Technology Analysis and Strategic Management*, **17** (4), pp. 445–76.

Geels, F. and J. Schot (2007), 'Typology of transition pathways in socio-technical systems', *Research Policy*, **36**, 399–417.

Göthenborg (2050), *Project Reports and Other Materials Related to the Gothenborg 2050 Project*, available at http://www.goteborg2050.se/ (accessed 19 March 2011).

Grin, J., J. Rotmans, J. Schot, J.F.W. Geels and D. Loorbach (2010), *Transitions to Change, Sustainable Development: New Directions in the Study of Long Term Transformative*, London: Routledge.

INRA-CIRAD (2009), *Agrimonde Scenarios and Challenges for Feeding the World in 2050*, Paris: NRA-CIRAD.

Jensen, J.D. and S. Smed (2007), 'Cost effective design of economic instruments in nutrition policy', *International Journal of Behavioural Nutrition and Physical Activity*, **4** (10), 1–12.

Kemp, R., F. Avelino and N. Bressers (2011), 'Transition management as a model for sustainable mobility', *European Transport*, **47**, 1–22.

Meadows, D.H., D.L. Meadows, J. Randers and W.W. Behrens (1972), *The Limits to Books: Growth. A Report for the Club of Rome's Project on the Predicament of Mankind*, New York: Universe.

Millenium Ecosystem Assessment (2005), *Ecosystems and Well-being: Scenarios*, Washington, DC: Island Press.

Mutombo, E. and T. Bauler (2009), 'Scenarios as sustainable development governance tools', available at http://www.belspo.be/belspo/ssd/science/Reports/A12_Mutombo_IHDP2009_Scenarios &SDgovernance.pdf (accessed 20 March 2011).

Pape, J. (2011), 'Sustainable food consumption in Ireland: challenges and opportunities', Presentation at the Lifestyles and Life-courses Workshop, University of Northampton, 20 April, available at http://www.consensus.ie/documents/perg_lifestyle_workshopjp_200411.pdf (accessed 20 September 2011).

Popkin, B.M. (2002), 'An overview on the nutrition transition and its health implications: the Bellagio Meeting', *Public Health Nutrition*, **5** (1A), 93–103.

Popkin, B.M. and P. Gordon-Larsen (2004), 'The nutrition transition: worldwide obesity dynamics and their determinants', *International Journal of Obesity*, **28** (3), 2–9.

Quist, J. (2009), 'Stakeholder and user involvement in backcasting and how this influences follow-up and spin-off', Paper presented at the Joint Actions on Climate Change Conference, Aalborg, Denmark, 8–10 June, available at https://gin.confex.com/gin/2009/webprogram/Paper2424.html (accessed 15 March 2011).

Quist, J. and P. Vergragt (2006), 'Past and future of backcasting: the shift to stakeholder participation and a proposal for a methodological framework', *futures*, **28** (9) 813–28.

Raskin, P., T. Banuri, G. Gallopín et al. (2002), *The Great Transition: The Promise and Lure of the Times Ahead*, Report of the Global Scenario Group, SEI PoleStar Series Report No. 10, Boston: Stockholm Environment Institute.

Reilly, M. and D. Willenbockel (2010), 'Managing uncertainty: a review of food system scenario analysis and modelling', *Philosophical Transactions of the Royal Society of London*, B, **365** (1554), 3049–63.

Reisch, L., A. Farsang and F. Jégou (2011), 'Scenario development for sustainable food consumption', CORPUS Discussion Paper 3, CORPUS – Enhancing the Connectivity between Research and Policymaking in Sustainable Consumption, funded by the European Commission, FP 7, Project No. 244103, available at http://www.scp-knowledge.eu (accessed 4 April 2011).

Reisch, L.A., U. Eberle and S. Lorek (2013), 'Sustainable for consumption: where do we stand today? An overview of issues and policies', *Sustainability: Science, Practice & Policy*, 9 (2), published online first.

Reisch, L., U. Eberle and S. Lorek (2013), 'Sustainable food consumption – issues and policies', Special Issue on Sustainable Food Consumption, *Sustainability*.

Reisig, V. and A. Hobbiss (2000), 'Food deserts and how to tackle them: a study of one city's approach', *Health Education Journal*, **59** (2), 137–49.

Rip, A. and R. Kemp (1998), 'Technological change', in S. Rayner and E.L. Malone (eds), *Human Choice and Climate Change*, Vol. 2, Columbus, OH: Battelle Press, pp. 327–99.

Rotmans, J. (1998) 'Methods for integrated assessment: the challenges and opportunities ahead', *Environmental Modelling and Assessment*, **3** (3), 155–79.

Rotmans, J. (2005), *Societal Innovation: Between Dream and Reality Lies Complexity*, Rotterdam: Erasmus University Rotterdam.

Rotmans, J., M. Van Asselt, C. Anastasi et al. (2000), 'Visions for a sustainable Europe', *Futures*, **32** (9), 809–31.

Rotmans, J., R. Kemp and M. Van Asselt (2001) 'More evolution than revolution: transition management in public policy', *Foresight*, **3**, 15–31.

Sedlacko, M. and N. Gjoksi (2010), *Future Studies in the Governance for Sustainable Development: Overview of Different Tools and their Contribution to Public Policy Making at European Sustainable Development Network*, ESDN Quarterly Report, March.

Shove, E. (2003), *Comfort, Cleanliness and Convenience: The Social Organization of Normality*, Oxford: Berg.

Slaughter, R.A. (1988), *Recovering the Future*. Clayton: Graduate School of Environmental Science, Monash University, Australia.

Slaughter, R. (1996), *New Thinking for a New Millennium*, London: Routledge.

Smith, A., A. Stirling and F. Berkhout (2005), 'The governance of sustainable socio-technical transitions', *Research Policy*, **34**, 1491–510.

Sondeijker, S. (2009), 'Imagining sustainability. Methodological building blocks for transition scenarios', PhD thesis, Erasmus University Rotterdam, Rotterdam.

Swart, R.J., P. Raskin and J. Robinson (2004), 'The problem of the future: sustainability science and scenario analysis', *Global Environmental Change*, **14**, 137–46.

Tapio, P. and O. Hietanen (2002), 'Epistemology and public policy: using a new typology to analyse the paradigm shift in Finnish transport futures studies', *Futures*, **34** (7), 597–620.

The Government Office for Science (2011a), *Foresight. The Future of Food and Farming: Final Project Report*, London: The Government Office for Science.

The Government Office for Science (2011b), *Foresight Project on Global Food and Farming Futures. Synthesis Report C1: Trends in Food Demand and Production*, London: The Government Office for Science.

The Royal Institute of International Affairs (2008), *Thinking about the Future of Food/The Chatham House Food Supply Scenarios*, Chatham House Food Supply Project.

UNECE (United Nations Economic Commission for Europe) (2012), *From Transition to Transformation: Sustainable and Inclusive Development in Europe and Central Asia*, New York: United Nations.

UNPD (United Nations Population Division) (2009), *2009 Revision of World Urbanization Prospects*, available at http://esa.un.org/unpd/wup/Documents/WUP2009_Highlights_Final.pdf (accessed 8 April 2011).

UNFPA (United Nations Population Fund) (2010), *State of World Population 2010. From Conflict and Crisis to Renewal: Generations of Change*, available at http://www.unfpa.org/webdav/site/global/shared/documents/publications/2010/EN_SOWP10.pdf (accessed 8 April 2011).

Van den Bergh, J. and F. Bruinsma (2008), *Managing the Transition to Renewable Energy: Theory and Practice from Local, Regional and Macro Perspectives*, Cheltenham, UK and Northampton, MA, USA: Edward Elgar.

Van der Heijden, K. (1996), *Scenarios: The Art of Strategic Conversation,* Chichester: John Wiley.

Van der Heijden, K., R. Bradfield, G. Burt, G. Cairns and G. Wright (2002), *The Sixth Sense: Accelerating Organisational Learning with Scenarios*, Chicester: Wiley and Sons.

Van Notten, P.W.F. (2005), 'Scenario development: a typology of approaches', Chapter 4 based on doctoral dissertation 'Writing on the wall. Scenario development in times of discontinuity', available at http://www.oecd.org/dataoecd/27/38/37246431.pdf (accessed 2 April 2011).

Van Notten, P.W.F., J. Rotmans, M.B.A. Van Asselt and D.S. Rothman (2003), 'An updated scenario typology', *Futures*, **35** (5), 423–43.

Verbong, G. and F. Geels (2007), 'The ongoing energy transition: lessons from a socio-technical, multilevel analysis of the Dutch electricity system (1960–2004)', *Energy Policy*, **35**, 1025–37.

Warde, A. (2005), 'Consumption and theories of practice', *Journal of Consumer Culture*, **5** (2), 131–53.

WHO (World Health Organization) (1996), *Malnutrition – The Global Picture*, World Health Organization, WHO Information fact sheets No. 119, available at https://apps.who.int/inf-fs/en/fact119.html (accessed 28 April 2011).

World Bank (2011), *Food Price Watch*, February, available at http://siteresources.worldbank.org/INTPREMNET/Resources/Food_Price_Watch_Feb_2011_Final_Version.pdf (accessed 21 April 2011).

PART II

Agency, behaviour and the European policy landscape

5. Emergent futures? Signposts to sustainable living in Europe and pathways to scale

Cheryl Hicks and Michael Kuhndt

INTRODUCTION

There is new momentum for change regarding the environmental and social impacts associated with our current European lifestyles. While many societal actors – from governments to business and civil society – may differ in their ideas about the scale of change required and what should be done, citizens from across Europe are taking action. There is now mounting evidence of social innovation, citizen and community movements that are driving notable change towards more sustainable ways of living in Europe. Signposts to future sustainable societies are emerging.

Individual examples of more sustainable living alternatives isolated to one part of our lives, one community or one demographic may appear to be niche exceptions to the mainstream of society. Taking a systemic view brings these examples together and provides us with a broader vision of what more sustainable living, and more sustainable societies, could look like. It is difficult for most people to imagine or visualize our lives in the future – and it is difficult to implement changes if we don't know where we're going. Visions and demonstrations of existing practice make abstract ideas become more tangible, and inspire action in order to realize our visions.

Promising practice, however, needs to scale up to mainstream market adoption if we are to achieve sustainable development goals in both the short and longer term. Technological innovation life-cycle and adoption models provide useful lessons by which to consider pathways to accelerate promising sustainable living practice to the mainstream. Evidence from the behavioural sciences on how behaviour change happens provides guidance on the importance of understanding differing individual needs, demands and desires according to their unique cultural and social contexts in order to achieve required change at scale.

This chapter examines the new evidence or groundswells of promising sustainable lifestyle alternatives emerging across Europe. It describes a vision for more sustainable ways of living, reviews possible drivers of citizen action and proposes a framework to analyse emerging trends towards more sustainable living. A number of promising practice examples are profiled according to the analytical framework, which seek to demonstrate that there is already a wealth of good practice to draw from, apply and modify from one context to others. The chapter argues that it will be important to continue to understand, test and evolve these promising concepts to ensure optimal positive impacts, and the consideration of possible unintended consequences or rebound effects (Hertwich, 2008). It reviews technological innovation models as well as lessons from the behavioural sciences to suggest possible pathways to scale up niche sustainable living practices to mainstream market adoption. The findings described in this chapter draw upon the recent work on sustainable consumption and behaviour change from the Collaborating Centre on Sustainable Consumption and Production (CSCP), the United Nations Government Programme (UNEP) Marrakech Process Task Force on Sustainable Lifestyles and the SPREAD Sustainable Lifestyles 2050: European Social Platform project (EU FP7) where the CSCP has been the coordinating partner.

SUSTAINABLE LIVING SIGNPOSTS AND DRIVERS

Our concept of more sustainable ways of living refers to those practices that have made progress towards the minimization of current harmful consumption and lifestyle impacts while optimizing quality of life and personal well-being. In order to address the current impacts of unsustainable European lifestyles and consumption patterns, we need to understand more specifically where the most significant impacts occur; what is driving those impacts; where the most significant improvements can be made; as well as the evidence of promising practice already underway demonstrating improvements (SPREAD, 2011).

What are Sustainable Lifestyles?

The most significant individual lifestyle impacts have been identified in the food we eat, the way we move around via transport and how we live in our homes in terms of energy and material use. These dominant individual lifestyle and private consumption impacts are referred to as impact 'hot spots' (CSCP, 2008). Food, mobility and housing account for 70–80 per cent of Europe's environmental impacts (Tukker et al., 2006). Signposts to more

BOX 5.1 ELEMENTS OF SOCIAL PRACTICE

Promising sustainable lifestyles practices address current lifestyle hot spots such as:

- Meat and dairy consumption: account for almost one quarter (24 per cent) of all final consumption impacts – by far the largest share in the food and drink sector (Weidema et al., 2008).
- Household energy use: domestic heating, water consumption, appliance and electronics use account for 40 per cent of Europe's total energy consumption (with space heating alone accounting for 67 per cent of household energy consumption in the EU-27) (EEA, 2010).
- Car ownership: related to dependency on single car use in the EU-27 increased by more than one third (35 per cent) between 1990 and 2007 (EEA, 2010). Over one third of the world's 750 million automobiles are owned by drivers in the European Union (EU) (IEA, 2010).

sustainable lifestyles refer to the promising practice examples that can demonstrate they are addressing one or more of these lifestyle hot spots.

Drivers of Change

In the past decade, there has been a range of factors coming together to catalyse citizen movements towards more sustainable living. People are proactively designing the communities they want to live in, and increasingly their aspirations for the 'good life' are supported by available infrastructure, products and services. From transition towns to eco-villages, and from 'slow food' to 'stay-cations', alternative transport services (car sharing, ride-sharing, high speed trains) and energy-efficiency upgrades in homes. This chapter explores these trends and their possible drivers.

The economic crisis of 2008 is often attributed to the renewed importance of cost savings, values of thrift, frugality and community due to the significant impact, or decline, in households' net financial wealth (Cussen et al., 2012). Market research reveals that consumers' values are evolving, so that values of conspicuous and hyper-consumption are more often questioned (Boston Consulting Group Study, 2011). Rising food, fuel and other commodity prices (World Bank, 2012) are leading to new quests for household efficiencies.

Growth in technological innovations that address the consumption and lifestyle hot spots (identified above) have enabled an increase in the availability and access to more eco-efficient products (such as household appliances, detergents, light bulbs and hybrid/electric cars) and services that are allowing us to consume differently by offering the benefit of a product, not the product itself, for example, carpet rentals, solar power, laundrettes (Botsman and Rogers, 2010; Manzini, 2002; Mont, 2009). Our increasing comfort with online social networking practices together with desires for closer community ties has developed into a proliferation of sharing, swapping and peer-to-peer lending services that has brought the value of ownership of goods into question in support of access to goods and services instead (Botsman and Rogers, 2010).

Growing desires for sufficiency or focus on quality of life and healthy living support the theory that the subjective well-being of people is linked to increasing income only up to a certain point. Once that point is reached, well-being and income decouple as increasing levels of income provide diminishing returns in terms of well-being and happiness (Armstrong, 2011; Max-Neef, 1995). The local-food movement has been gaining momentum with consumers seeking alternatives to the industrial agriculture system or rising concerns over the impacts of transport linked to globalization (Nielsen, 2012; World Watch Institute, 2010). Europeans are also getting involved in individual and community food production via urban gardens (Schausberger, 2012).

Examination of policy instruments and initiatives in Europe, which could be linked to recent trends towards more sustainable ways of living, has found evidence of a broad range of policy instruments in a number of jurisdictions. The CSCP's policy instrument classification system has been used to consider policy initiatives beyond traditional regulatory or 'command and control' approaches, and differentiated by whether the policy provides assistance and support, motivation or rewards for producers and consumers: regulatory, economic, informational, cooperation and educational instruments (CSCP, 2010).

SUSTAINABLE LIVING GROUNDSWELLS FROM ACROSS EUROPE: EFFICIENT, DIFFERENT AND SUFFICIENT

According to current trends, and the drivers of change outlined above, lifestyles of the future could be more efficient, different and sufficient.

Numerous examples of promising sustainable living examples have been identified from across Europe that are demonstrating improvements against unsustainable consumption and lifestyle hot spots – in the areas of food, mobility and housing. These promising practices can be categorized into three dominant trend areas: efficient lifestyles, different lifestyles and sufficient lifestyles (JCP, 2008; Speth, 2008; UNEP, 2001, 2009; SPREAD, 2011; WWF, 2011). This chapter proposes these trend areas as a framework to analyse promising and emergent trends towards more sustainable living.

Efficient Lifestyles

This category refers to current household behaviour trends towards wasting less, including the choice, use and disposal of more efficient products and services. Efficient living reduces critical lifestyle impacts such as overconsumption, waste and the resulting exploitation of natural resources (Cooper, 2002) – while often delivering cost savings to the individual. There is new evidence that Europeans have been wasting less energy, water, food and materials in recent years. A recent Eurobarometer survey showed an increase in the number of people in the EU-27 reducing energy and water consumption in addition to reducing and separating household waste (EC, 2011). This subsection focuses its analysis of sufficient lifestyles on the lifestyle consumption hot spots of food, and the emegent trends in community design movements.

A 2009 survey of over 3000 homeowners in five EU countries demonstrated that almost half of EU homeowners are investing in energy savings for the home, and that a quarter had installed a new boiler/heating system and had fitted their homes with double or triple-glazing (Tuominen and Klobut, 2009). The energy efficiency of heating and large electrical appliances has, on average, increased by 1 per cent and 1.5 per cent, respectively over the past decade. The market share of EU A-label appliances (indicating that they are the most energy efficient) grew from just 10 per cent to as high as 90 per cent between 1998 and 2008 (European Commission, 2011). Decentralized renewable energy production has also grown significantly in private households. The diffusion of solar water heaters, for example, reached 75 per cent of households in Cyprus in 2000, with Greece also making significant progress from 2000 to 2009 (35 per cent of all dwellings equipped with a solar water heater). In Austria, 24 per cent of dwellings have such heaters, 11 per cent in Malta and 6 per cent in Germany (Odyssee/Enerdata, 2011). A common low energy standard is the so-called 'passive' house that requires total energy demand for space heating and cooling be capped at 15 kWh/per square metre (PEP, 2011).

By 2007, over 8000 passive houses had been built in Europe (Passive-On, 2011).

Policy designs linked to efficient living include energy ratings for homes, incentives for home efficiency improvements, promotion of eco-efficient products and services and efficiency and waste management awareness programmes. Building efficiency standards have been an attractive economic and political option to policy makers because of the significantly large impact of buildings in terms of total energy usage. As part of the EU's Directive on The Energy Performance of Buildings, since 2009, all member countries are expected to have energy performance certificates available as part of a required information packet when a new home is sold or leased. Most (if not all) EU countries have financial support schemes for energy-efficiency renovations. However, there are significant differences in the existing housing stock, regulations for renovation of existing buildings and new construction and different financial resources available to building owners in different countries across Europe. In Spain, the Technical Building Code has, since 2006, included an obligation to meet up to 70 per cent of domestic demand for hot water through solar thermal energy (for new buildings and major renovations). In the UK, a joint venture between a local energy authority, Kirklees Council and Scottish Power offered free cavity wall and loft insulation to citizens from 2007–10. From 2012, the UK's new Behavioural Insights Unit will issue redesigned Energy Performance Certificates (EPCs) that will tell people how costly it will be to heat a home they are buying. It is hoped that this will 'nudge' 1.4 million households to make their homes more energy efficient, saving them money in the process (McKinsey Quarterly, 2011). Energy savings of up to 20 per cent have been estimated to be possible by 2030 through the renovation of private housing stock in Europe alone (Tuominen and Klobut, 2009).

Promising practice in energy efficiency does show progress, but we have not yet reached the tipping point of mass-market adoption. In its 2010 report, the EEA reported a 30 per cent increase in household electricity consumption between 1990 and 2007 despite rising energy costs and recent gains in energy efficiency in buildings and appliances attributed to the steady increases in the numbers of appliances, electronics and communication equipment acquired by European households. Improvements to the current incentive programmes are still required in terms of information, application procedures and scope (that is, including people who have limited finances). Homeowners do appreciate and make ample use of available schemes (Adjei et al., 2010; Bartiaux, 2011). Lessons from technological innovation models and the behavioural sciences are explored later in the chapter.

Different Lifestyles

This group refers to a shift in preferences from quantity and ownership to access and quality, a movement also called 'collaborative consumption' (Botsman and Rogers, 2010) – a shift in how we live, move and consume. It also includes the experience economy (Hirschman and Holbrock, 1982; Pine and Gilmore, 1998), referring to the growing importance of the emotional experiences linked to products and services rather than the products themselves, and product service systems (PSS) (Manzini and Vezzoli, 2002), a customized mix of services as a substitute for the purchase and use of products in order to provide a specific integrated solution to meet a customer's needs or desires. Different living offers the potential to decouple material consumption from resource use.

An influence of different lifestyle trends on the consumption and lifestyle hot spot of single car use dependency provides an example of this. There are indications that people are changing the way they move around in favour of alternative access to mobility versus single car use and ownership; they are increasingly choosing different alternatives. The use of high-speed trains are amongst the transport modes that have recently seen the highest growth rates, with passenger kilometres increasing by 180 per cent in the EU-27 between 1995 and 2007 (EC, 2009). In some cases, such as in France, the expansion of the high-speed train network has been successful in reducing the demand for air travel (EEA, 2010). While there are differing views as to the total life cycle environmental impacts of new high-speed train infrastructure (WWF, 2008), it is the evidence of behaviour change that makes this practice promising and different.

Sharing, swapping and lending have seen a dramatic growth in the latter part of the last decade demonstrating alternatives to how we move around and consume goods. Europeans are sharing cars, bikes, houses and household goods. By 2015, it is estimated that 5.5 million people in Europe will belong to car-sharing services like Cambio Mobilitätsservice and DB Rent Flinkster in Germany, or Mobility Carsharing in Switzerland. A 2010 study by the European project MOMO on car sharing assessed station-based car-sharing services in 14 European countries and reported approximately 380 000 car-sharing participants having access to approximately 11 900 car-sharing vehicles (latest figures 2009). UK-based WhipCar, a neighbour-to-neighbour car-sharing service, had over 1000 owners accepting bookings within the first six months of launch (Botsman and Rogers, 2010). In Europe, bike sharing also increased in 2007, when the number of public bicycle schemes doubled (Spycycles, 2009) – a trend that can now be observed in most major European cities.

Policies linked to 'different' living include incentives for reduced car use, consumer empowerment and education. While many of the reforms are based around standard economic instrument tools like taxing car use and regulating available parking spaces, some cities have also used innovative physical choice architecture (McKinsey Quarterley, 2011). As an example of choice architecture, in Stockholm sections of curbside have been marked with one large box, sometimes taking up an entire street, therefore limiting space for parking and encouraging citizens to leave their car at home. Campaigns and initiatives to support cycling or taking the bus to school and work have shown that this is a viable and sustainable alternative for mobility from an environmental, economic and social point of view. In France, the local authorities of Paris have launched one of the biggest bike share programmes in Europe, *The Velib*, placing 20 000 bikes and more than 1000 stations around Paris. These cities demonstrate that congestion charges, limiting parking places and pedestrian zones working in combination can create the conditions to significantly reduce car use (EC, 2004).

Sufficient Lifestyles

This group categorization refers to the focus on improving quality of life while making conscious efforts to consume less. As our lives become more complex, a growing number of people prefer to shape their life in a simpler way to reduce the pressure created by an over abundance of 'stuff' or to reduce the adverse impacts of overconsumption. The 'voluntary simplicity movement', where people leave their business-as-usual jobs to focus on local, small-scale activities, is also gaining momentum in Western Europe (Alexander, 2011). New low impact lifestyle trends are emerging including the Slow Living movement consumerism and voluntary downshifting (McDonald et al., 2006). LOVOS refers to 'lifestyles of voluntary simplicity', which is oriented to health and sustainability and is critical of consumption. There is a resurgence of public campaigns against materialistic ways of life such as the 'Buy Nothing Day' campaign. Concerns about materialistic lifestyles amongst the younger generations and a high environmental and social awareness are all positive trends that suggest that alternative lifestyles have the potential to proliferate in the future. This subsection focuses analysis of sufficient lifestyles on the lifestyle consumption hot spots of food, and the emergent trends in community design movements.

In relation to food, Europeans have been consuming less meat in recent years. This trend accelerated in 2008, often attributed to increases in food prices, when total meat consumption dropped by 2.2 per cent in the EU-27 compared to 2007 (EC, 2009; EEA, 2010). A major reduction in meat and dairy consumption (accounting for 50 per cent of the ecological footprint of our food), generated by eating local, seasonal food, and through improved farming efficiency and reduced food waste could cut the ecological footprint of food by as much as 60 per cent (Bio-Regional and CABE, 2008). Evidence is not conclusive in terms of the drivers of this data on the voluntary reduction of meat consumption. However, in addition to increases in food prices, there have been increasing trends in education and awareness campaigns across Europe such as the Weekly Veggie Day, which encourages citizens to go one day a week without meat or fish (Wikipedia, 2012).

There has also been a trend towards voluntary community redesign around sufficient lifestyles, not least through the Transition Towns model that emerged from the works of Bill Mollison (1988) and David Holmgren (2003) and more recently through the work and campaigning of designer Rob Hopkins and his students (2005). Transition Towns work with existing community groups, local authorities and interested individuals to set up locally based solutions in a variety of areas. These include promoting practical skills and training programmes, establishing land allotments, working with local businesses, improving energy efficiency and establishing local currencies (LETS) (Mont and Power, 2010). The initiative has spread rapidly, and as of September 2011, there are almost 400 communities recognized as official Transition Towns in approximately 34 countries according to the movement's official website. Eco-villages such as BioRegional's global One Planet Communities (OPC) initiative consists of a range of practical projects and partnerships that demonstrate how we can live a healthier, high-quality lifestyle within equitable share of the Earth's resources. In the development of a one planet community, members work in partnership with municipal authorities, local residents, voluntary organizations and businesses to put in place initiatives that can be adopted throughout the area to help organizations and individuals to reduce their environmental footprint. The London Borough of Sutton and Middlesbrough were the first OPCs. Today, OPCs span the USA, UK and Portugal, with others planned in South Africa, China, Australia and Canada.

Individual families are also experimenting with self-sufficient living. An average family uses 4000 kWh of energy a year. Low Impact Man Steven Vroman from Ghent, Belgium, aims to make his family's ecological footprint as small as possible: 1 kWh a day. He bought a new fridge, washes full loads and threw out the microwave and iron. In Sweden, the Lindell family was selected to demonstrate how to use less than 80 kg of carbon dioxide (CO_2) per week as part of the One Tonne Life project. The family lived in a new wooden house with energy-efficient windows, roof and facade-integrated thin-film solar panels, a recycling station, whole-house energy monitoring, light-emitting diode (LED) lights low-flow fixtures, and walls covered with clean, white wood panels. The Lindells also used a sporty Volvo C30 EV that's charged with power captured from solar panels and configured to the carport. The One Tonne Life was meant to be as much about decisions as the things that will enable this four-person family to live a carbon-smart year. Everything from food choices to the mode of transportation was considered.

Policies and public initiatives linked to sufficient living include the promotion of eating a more balanced diet – as opposed to overconsumption of meat – as well as provisions for investment in eco-towns and co-housing legislation. The UK government's collaborative food procurement programme was established to identify how public sector organizations can work together when buying food to achieve improved sustainability and different diets. The Welsh Community Food Co-operative Programmes provide quality, affordable fruit and vegetables to communities through sustainable local food distribution networks. The 'Roots and Shoots' programme, in selected schools in England, teaches children how to plant and grow food, and to join 'growing clubs'. In Belgium, the Flemish government provides schools with supplies of fruit and vegetables with the aim of motivating children to learn to eat more fruit and vegetables as opposed to meat. Food policy in Sweden has been targeted at sustainable food production and to increase public knowledge on the connections between diet and health. The Think Twice programme, with a four-year action plan (2005–09), aimed to teach consumers in Sweden about healthy eating, living and travel. In Romania and Denmark, 'fat taxes' aimed at fast and processed food have been provisioned to help consumers rethink the way they eat.

European municipalities are increasingly playing a leading part in the development of eco-towns across Europe. Local authorities are getting involved in the master planning process to help achieve consensus, avoid duplication and reduce the risk to private sector participants (PRP Architects Ltd URBED and Design for Homes, 2008). Public financial institutions often supply long-term debt finance at low rates of interest for installing infrastructure, to be repaid from land sales, rather than relying on a 'lottery' of grants or government patronage. Public finance for transport and environmental infrastructure are frequently funded upfront, either by the local authority or by utility companies, using low cost, long-term finance with the investment recovered over a longer period. Land is often assembled by the public sector, and paid for as sites are sold or homes are occupied. Examples of eco-towns where local authorities have played a leading role are Adamstown, near Dublin; Amersfoort, the

Netherlands, with its three new suburbs: Kattenbroek, Nieuwland and Vathorst; Freiburg, Germany, with its two new urban extensions: Vauban and Rieselfeld; HafenCity in Hamburg, Germany; Kronsberg in Hanover, Germany; and Hammarby Sjo stad, an urban extension of Stockholm, Sweden.

Despite these promising trends, however, there remains little evidence or data in relation to the ways in which these initiatives are influencing either relative decoupling (reduced environmental impact per unit of gross domestic product (GDP)) or progress towards absolute decoupling (reduced overall environmental impact) (Jackson, 2009). It is clear, therefore, that further research, monitoring and testing of these promising practice examples will be essential to developing pathways to scale and the market adoption that will deliver the necessary changes that will be needed to bring about more sustainable European lifestyles.

PATHWAYS TO SCALE: LESSONS FROM TECHNOLOGICAL INNOVATION LIFE CYCLE AND MARKET ADOPTION MODELS

With the emergence of promising sustainable living alternatives, questions remain on how to they might be scaled up into a more mainstream market approach. To envision what pathways to scale for sustainable living might look like, this section reviews and assesses relevant examples of market adoption strategies from industry. The study of patterns in the evolution of technological innovation has given guidance to managers regarding different business strategies to deploy at different stages of a product, or industry's, maturity. These studies have also offered insights into what types of companies, or innovations, are most likely to succeed or fail in a range of circumstances (Christensen, 2000). This chapter argues that the study of patterns in the evolution of technological innovation could also give valuable guidance on possible strategies to deploy to scale up more sustainable living practice in Europe.

The 'diffusion of innovations' model suggests that market adoption reflects a bell curve that tracks to customer/consumer adoption of a new technology, product or service (Moore, 1990; Rogers, 1962). The bell starts with early adopters, interested in trying something new. It then moves to early majority target markets, which may require some customization of the product, and become reference points for other segments. Mass-market adoption then moves from customization to mass-market manufacturing and distribution. From there the market matures and late adopters and risk-adverse laggards adopt the 'tried and true' solutions.

The S-curve theory is a popularized innovation theory that suggests that the pace of technological progress (product, service, technology or business progress) follows an 'S' curve pattern as it evolves over time (Foster, 1986). S-curve theory suggests that the rate of technological progress is ultimately subject to decreasing returns because progress is eventually constrained by natural limits of some sort (Sahal, 1981). When the rate of technological progress has begun to decline, the technology is vulnerable to being overtaken by a new technological approach. Usually the end of one S-curve marks the emergence of a new S-curve – the one that displaces it. A popularized example of the evolution of a technology along the S-curve and the emergence of new S-curves is that of the technological innovation evolution from video-cassettes to CDs and now MP3 and from typewriters to word processors and then computers.

In the market adoption bell curve, competitiveness is based on incremental improvements and economies of scale. Incremental innovation thrives in structured environments characterized by continuous product and process improvements. Incremental innovation may sustain steady growth within an S-curve. However jumping curves, and maintaining market momentum, involves creating and driving disruptive innovations – creating new to the world customer value (such as mobile phones or tablets), and displacing existing ways of delivering value (such as digital versus film photography). A key management strategy has been to monitor a company's position on its S-curve and when it is approaching market adoption to develop the new technology that might overtake the present approach to avoid obsolescence (Christensen, 1992). Some industries and technologies move along the S-curve faster than others. High-tech industry S-curves tend to cycle more quickly. These industries have found that it is also important to 'manage the edges' of the bell curve, the early adopters and emergent majority (Gregory, 2010).

By understanding where current groundswells of more sustainable living fall on the S-curve innovation life cycle, strategies can be deployed to avoid obsolescence, or rebound, and to instead leverage this knowledge to accelerate scale. 'Efficient living' in the context of the S-curve methodology can be considered incremental innovation – continuous product and process improvements that lead to greater efficiency and benefit to the consumer. However, this does not create 'new to the world customer' value as in the case of disruptive innovation. Many early examples of efficient living in Europe, it can be argued, have moved along the bell curve from early adopters to the emergent majority; such as in energy-efficient appliances, for example, where market share growth has gone from 10 to 90 per cent in a decade. In this case, to achieve scale – according to the technological innovation life cycle – energy-efficient appliance companies should focus on meeting current customer needs to continue to grow and maintain market share, while also thinking about the next innovation S-curve that will displace it or what will displace energy-efficient appliances. This thinking might also be applied to policies. For example, if energy-efficiency policy incentives for homes are reaching the emergent majority in Europe, what should come next?

'Different living' in the context of the S-curve methodology could be considered disruptive innovation or jumping curves, creating new to the world value. Different living practices, for example, the sharing economy, have the potential for growth spurts, behaviour change and change in social norms. At this stage, the bell curve theory suggests that it is important to understand the early adopter and emergent majority segments in order to apply the customization that will attract new segments and provide the transition to the mass market. For example, what customization can be applied to car-sharing and bike-sharing services to attract new segments? Where can policy support disruptive innovation?

'Sufficient living' in the context of the S-curve methodology can be considered emerging practice or the start of a new curve. The management of emerging practice suggests a focus on anticipating future individual needs (exploring uncertainties), rewarding experimentation, allowing freedom and flexibility and putting mechanisms in place to monitor and build on learning. Beyond its current state of experimentation, to scale up, bell curve methodology suggests that the next step for niche eco-communities, for example, will be to customize future development in order to attract new segments. After an analysis of future individual needs of existing users, eco-community developers could explore the needs of new segments in order to customize new offerings.

PATHWAYS TO SCALE: LESSONS FROM THE BEHAVIOURAL SCIENCES

Equally important for market adoption of more sustainable living practices is a deeper understanding of people and behaviour – what motivates people to act. Individual lifestyle choices can be shaped not only by cultural norms, personal desires and societal trends, but also by national policies, availability and accessibility of resources, goods and services (Breukers et al., 2009). Behaviour change can be dependent on a person's sense of agency and in today's society is increasingly driven by one's need for instant gratification, or one's ability to delay it. Old habits die hard. For a variety of different reasons, people are locked into unsustainable patterns of behaviour. Interruption or intervention is required to initiate change, support and reinforcement to maintain the change. Change can come from the individual, from our social groupings and shared value systems or from influencers and interaction with our environment (Hicks and Hovenden, 2010).

To interrupt current unsustainable lifestyle lock-ins, and to motivate people to change behaviour, we have to make alternative and more sustainable solutions easy (Thaler and Sunstein, 2008), accessible and convenient with minimum compromise on quality or price (WBCSD, 2008). Change, for individuals and groups, often occurs when significant change or change events undermine and demand a reformation of the value system. There is evidence that it is possible to change behaviours, but old behaviours have to be unlearned first. When behaviours are strongly associated with reward and pleasure, as is the case for most of our current consumption, the unlearning is very difficult because strongly reinforced neuronal connections in the brain have to be broken (Doidge, 2007). For change to occur without trauma, the behavioural sciences suggest that several key things have to happen: changes proposed have to satisfy the individual's needs; old behaviours need to be unlearned; unknown changes need space; positive reinforcement or feedback is critical; and focusing on one change at a time produces lasting results (Hicks and Hovenden, 2010). In order to suggest possible pathways to scale current sustainable living trends, this section applies these lessons from the behavioural sciences to our framework of efficient, different and sufficient living.

Changes proposed have to fill the individual's needs. In the case of efficient living, efficiency changes will only be adopted if they maintain or ideally enhance perceived quality of life elements such as those things in one's life that bring feelings of pleasure or success. A drop in performance or perceived status will not be accepted. There will be broad differences in aspirations for 'the good life' (the elements that individuals define as pleasure or success for themselves) across household segments. Therefore, many efficiency options will need to be available and accessible to fulfil differing individual needs.

Old behaviours need to be unlearned. In the case of different living, shifts in preferences from ownership to access to goods and services or the experience economy require that the tendencies associated with ownership need to be unlearned. If it is clear that the access or experience of that product or service is clearly more desirable, pleasurable or fulfils a specific need, then the unlearning of old behaviours is less of an issue. In the case of mobility, this is why driving a hybrid will be easier for most Western consumers than beginning to bicycle, or take public transport, because the same need is fulfilled. However, where new behaviours are needed (such as following the fixed timetable of public transport) they work best when phased in so that new behaviours are learned in a non-critical way and old behaviours are gradually reduced. Car sharing would be one example of the step change to unlearning single car dependency and towards public transport use.

Unknown changes need space. In the case of sufficient living, behaviour change options relevant to differing individual preferences are often not yet known, therefore people need to make space for change. This requires a difficult transition, since old behaviours tend to be very persistent, and suggests why things like support groups, retreats or regular meetings are often required. In the case of food sufficiency (or consuming less meat), health and culinary education and awareness building are often required.

Positive reinforcement or feedback is critical. This allows people to keep connecting change to things that are important to them. In the case of energy efficiency, continuous feedback on energy use in the home or in your car has been shown to change behaviour (Thaler and Sunstein, 2008).

CONCLUSION: A FRAMEWORK FOR ACTION

Evidence suggesting the minimization of harmful environmental and social impacts associated with our current lifestyle patterns in Europe is not conclusive. Nor is there yet evidence of change in mass-market consumption patterns that would lead to absolute decoupling. This chapter, however, has focused on the argument that there is new momentum for change amongst European citizens, through numerous examples of promising practice of changes to lifestyle patterns in new ways that may be seen as signposts to emergent futures where improvements will become more evident. Organizing these promising practices into a framework of trend areas – efficient, different and sufficient living – has aimed to provide a vision of what more sustainable living would look like, and to assist in the recognition of how these trends could move along the market adoption bell curve towards mass-market adoption in order to foster, facilitate and accelerate the urgent need for scale up and progress. The chapter has argued that the framework is equally relevant in identifying strategies for behaviour change for different trend areas and segments based on lessons from the behavioural sciences.

Efficient living trends are showing that Europeans have been wasting less energy, water, food and materials in recent years. This chapter has examined efficient living in relation to addressing the lifestyle impact hot spot of energy use in the home. It has found that Europeans are investing in energy savings for the home such as insulation, energy-efficient appliances and caps on space heating. These trends are being driven by changes in net household wealth, commodity prices and growth in access to energy-efficient products and services. Policy incentives have been particularly impactful in the areas of energy ratings for homes, incentives for home-efficiency improvements, promotion of eco-efficient products and services, efficiency and waste management awareness programmes.

Different living trends are showing the potential to decouple material consumption from resource use through shifts in preferences from quantity and ownership of products to access and quality. The chapter has focused on different living trends addressing the lifestyle hot spot of single car use dependency and has found groundswells of citizen movements gravitating towards car sharing, bike sharing and increased use of high-speed trains and public transport systems. Policy incentives for reduced car use have included taxing car use, regulating available parking spaces, consumer empowerment and education, as well as choice architecture and nudging.

Trends towards sufficient living are showing a growing number of people preferring to shape their lives in a simpler way to reduce the pressure created by an overabundance of 'stuff' or to reduce the adverse impacts of overconsumption. Growth in Transition Towns, eco-villages and movements towards local food consumption and community or urban gardening are examples of sufficient living trends that this chapter has examined. Policy incentives have been identified in the areas of healthy, balanced diet promotion – addressing the lifestyle hot spot of overconsumption of meat and dairy – as well as provisions for investment in eco-towns and co-housing.

Policy, industry and civil society can play important roles in fostering and accelerating the pathways to scale sustainable lifestyles to mass-market adoption. Resilient change requires enabling environments and infrastructures to support long-lasting behaviour change (SPREAD, 2011). Policy incentives and investments in the mass-market adoption of current promising sustainable living innovations – products, services and social innovation – can be important drivers for change.

The chapter has also argued that the study of patterns in the evolution of technological innovation could give valuable guidance on the strategies to deploy to scale up promising sustainable living practices in Europe. By understanding where current groundswells of more sustainable living fall on the S-curve innovation life cycle, strategies can be deployed to avoid obsolescence, or rebound, and to instead leverage this knowledge to accelerate scale and influence. Efficient living in the context of the S-curve methodology can be considered incremental innovation – continuous product and process improvements that lead to greater efficiency and benefit to the consumer; different living could be considered disruptive innovation or jumping curves, creating new to the world value; and sufficient living can be considered emerging practice or the start of a new curve.

Just as technological innovation will not be enough to bring current lifestyles and our consumption patterns to sustainable levels, lessons from technological innovation market adoption will also not be enough. The chapter has therefore argued that lessons from the behavioural sciences on how behaviour change happens will also be required in order to scale promising behaviour changes associated with alternative and more sustainable living trends. The chapter has also applied examples of efficient, different and sufficient living to lessons from the behavioural sciences to suggest possible pathways to scale such as: satisfying individual needs; unlearning old behaviours; providing changes with needed space; providing positive reinforcement or feedback; and focusing on one change at a time to produce lasting results.

Opportunities for Action

Governments can learn from the policy examples that are already showing promising results across Europe in achieving lasting behaviour change models. They can learn from technological innovation market adoption models and lessons from the behavioural sciences to create policy incentives that foster and accelerate scale up and mass-market adoption.

Large businesses can leverage their expertise in understanding technological innovation market adoption models and understanding of consumer household needs and desires to focus product and service innovation enabling more sustainable living that is easier and accessible for diverse household contexts. They can invest in the small sustainable businesses and entrepreneurs that will be needed to achieve 100 per cent sustainably sourced supply chains.

Entrepreneurs and entrepreneurship networks can consider the key impact areas of current lifestyles (food, mobility, housing) as new business opportunity spaces for the business ideas that can address them.

Investors can prioritize investments in businesses and infrastructure that address current lifestyle impact hot spot areas of food, mobility and housing that will deliver both sustainability impact returns as well as financial returns. Positive policy activity, increasing consumer demand, opportunity for scale, replicability and evolved understanding of physical risks all favour a positive investment case.

Educators can become active in designing the future of learning that includes the life skills that will be needed to maintain and thrive in new sustainable living contexts – and to support future sustainable societies.

Researchers can continue to track, monitor and analyse the promising sustainable living trends emerging in order to develop our understanding of which practices and enablers actually make people better off and deliver the future lifestyles we want. The chapter has sought to express optimism for the prospects of a future of more sustainable living for all. Our challenge lies in seizing the opportunities fast enough to bring about resilient change.

REFERENCES

Adjei, A., L. Hamilton and M. Roys (2010), *A Study of Homeowners' Energy Efficiency Improvements and the Impact of the Energy Performance Certificate*, Deliverable 5.2 of the IDEAL EPBD project, supported by Intelligent Energy Europe, available at http://ideal-epbd.eu/download/homeowners_questionnaire_wa.pdf (accessed 6, September, 2012).

Alexander, S. (2011), 'The voluntary simplicity movement: reimagining the good life beyond consumer culture', *International Journal of Environmental, Cultural, Economic and Social Sustainability*, **7** (3), 133–50.

Armstrong, A. (2011), 'Mindfulness and compulsive buying', Research Group on Lifestyles, Values and Environment end of award Conference Living Sustainably: Values, Policies, Practices, London, 15 June.

Bartiaux, F. (2011), *A Qualitative Study of Home Energy-related Renovation in Five European Countries: Homeowners' Practices and Opinions*, Deliverable 4.2 of the IDEAL EPBD project, supported by Intelligent Energy Europe, available at http://ideal-epbd.eu/download/in_depth_interviews.pdf (accessed 6, September, 2012).

Bio-Regional and CABE (2008), *What Makes an Eco-town?*, London: BioRegional Development Group and the Commission for Architecture and the Built Environment (CABE), p. 32.

Boston Consulting Group (2011), *Luxe Redux: Raising the Bar for the Selling of Luxuries.*

Botsman, R. and R. Rogers (2010), *What's Mine Is Yours: The Rise of Collaborative Consumption*, London: HarperCollins.

Breukers, S., E. Heiskanen, R.M. Mourik et al. (2009), *Interaction Schemes for Successful Demand-side Management. Building Blocks for a Practicable and Conceptual Framework*, Deliverable 5 of the Changing Behaviour project, supported by the European Commission, available at http://energychange.info/deliverables/191-d5-interaction-schemes-for-successful-energy-demand-side-management (accessed 6 September, 2012).

Christensen, C.M. (1992), 'Exploring the limits of the technology S-curve, parts I and II', *Production and Operations Management*, **1** (4), 358–66.

Christensen, C.M. (2000), 'The evolution of innovation', in R. Dorf (ed.), *Technology Management Handbook*, Boca Raton, FL: CRC Press, pp. 2–11.

Cooper, T. (2002), 'Durable consumption: reflections on product life cycles and the throwaway society', in *Workshop Proceedings: Life-cycle Approaches to Sustainable Consumption*, 22 November, Vienna: IIASA, pp. 15–27.

CSCP (2010), 'Sustainable consumption and production policies – the role of civil society organisations', *A Guide for Civil Society Organisations*, published in the framework of the Action Town project.

Cussen, M., B., O'Leary and D. Smith (2012), 'The Impact of the financial turmoil on households: a cross country comparison', *Quarterly Bulletin*, 2 April.

Doidge, N. (2007), *The Brain that Changes Itself*, New York: Penguin.

EC (European Commission) (2004), *Reclaiming City Streets for People – Chaos or Quality of Life?*', Luxembourg: Office for Official Publications of the European Communities.

EC (European Commission) (2009), *Communication from the Commission – A Sustainable Future for Transport: Towards an Integrated, Technology-led and User friendly System*, COM(2009) 279 final available at http://eur-lex.europa.eu/LexUriServ/LexUriServ.do?uri=COM:2009:0279:FIN:EN:PDF (accessed 6 September, 2012).

EC (European Commission) (2011), *Roadmap to a Resource Efficient Europe* COM(2011) 571 final available at http://ec.europa.eu/environment/resource_efficiency/pdf/com2011_576.pdf.

EEA (2010), 'The European environment – state and outlook 2010: consumption and the environment', European Environment Agency, Copenhagen.

Foster, R.N. (1986), *Innovation: The Attacker's Advantage*, New York: Summit Books.

Gregory, E. (2010) 'Innovation for the common good: leading the way. What public agencies, philanthropists, private companies and civil society can do to invent the future together', Orange Paper No. 1, Collective Invention, San Francisco.

Hertwich, E.G. (2008), 'Consumption and the rebound effect: an industrial ecology perspective', *Journal of Industrial Ecology*, **9** (11–2), 85–98.

Hicks, C. and F. Hovenden (2010), 'Making sustainability easy for consumers: the new opportunity for business and societal innovation' in G. Williams (ed.), *Responsible Management in Asia: Perspectives on CSR*, Basingstoke, Palgrave: Macmillan, pp. 133–45.

Hirschman, E. and M. Holbrook (1982), 'Hedonic consumption: emerging concepts, methods and propositions', *Journal of Marketing*, **46**, (3, Summer).

Holmgren, D. (2002), *Permaculture: Principles and Pathways beyond Sustainability*, Hepburn, Vic: Holmgren Design Services

Hopkins, R. (2008), *The Transition Handbook: From Oil Dependency to Local Resilience*, Dartington: Green Books Ltd.

IEA (2010), *Energy Technologies Perspectives*, Paris: OECD/IEA, available at www. iea.org/publications/free publications/publication/etp 2010.pdf.

Jackson, T. (2009), *Prosperity Without Growth*, London: Earthscan.

JCP (2008), 'Fostering change to sustainable consumption and production: an evidence based view', *Journal of Cleaner Production*, **16**, 1218e1225.

Manzini and Vezzoli (2002), *Product Service Systems and Sustainability: Opportunities for Sustainable Solutions*, Paris: UNEP, CIR.IS and IIIEE Lund University.

Max-Neef, M. (1995), 'Economic growth and quality of life: a threshold hypothesis', *Ecological Economics*, **15**, 115–18.

McDonald, S., C.J. Oates et al. (2006), 'Toward sustainable consumption: researching voluntary simplifiers', *Psychology and Marketing*, **23** (6),: 515–34.

McKinsey Quarterly (2011), 'Nudging the world toward smarter public policy: an interview with Richard Thaler', *McKinsey & Company*, June.

Meroni, A. (2007), *Creative Communities: People Inventing Sustainable Ways of Living*, Milan: Edizioni POLI.design.
Mont, O. and K. Power (2009), 'Understanding factors that shape consumption', unpublished manuscript, Copenhagen, ETC-SCP and EEA, p. 103.
Mollison, B. (1988), *Permaculture: A Designer's Manual*, Tyalgum, NSW: Tagari Publications.
Mont, O. and K. Power (2010), 'The role of formal and informal forces in shaping consumption and implications for a sustainable society. Part I', *Sustainability*, **2** (5), 2232–52.
Moore, G. (1990), *Crossing the Chasm*, New York: Harper-Collins.
Nielsen (2012), *Global Trends in Healthy Eating*, The Nielsen Company, http://blog.nielsen.com/nielsenwire/consumer/global-trends-in-healthy-eating/ (accessed 6 September, 2012).
Odyssee/Enerdata (2011), *Energy Efficiency Trends for the Households in the EU-27*, Supported by Intelligent Energy Europe, available at http://www.odyssee-indicators.org/reports/ee_households.php (accessed 16 September, 2011).
Passive-On (2011), *The Passive-On Project: The Passivhaus Standard*, available at http://www.passive-on.org/en/details.php (accessed 17 September, 2011).
PEP (2011) *Promotion of European Passive Houses: Passive House Definition*, available at http://pep.ecn.nl/what-is-a-passive-house/ (accessed 17 September, 2011).
Pine, B. and J. Gilmore (1998), 'Welcome to the experience economy', *Harvard Business Review*, July.
PRP Architects Ltd, URBED and Design for Homes (2008), *Beyond Eco-towns: Applying the Lessons from Europe*, London: PRP Architects Ltd.
Rogers, E. (1962), *Diffusion of Innovations*, New York: The Free Press, a division of Simon & Schuster.
Schausberger, B. (2012), 'Assessing a trend: community gardens are on the rise across European cities' *Urban Gardening*, 2 April.
Siegel, C. (2010), 'Livable cities and political choices', available at http://www.planetizen.com/node/44299 (accessed 20 May, 2010).
Speth, J.G. (2008), *The Bridge at the Edge of the World: Capitalism, the Environment, and Crossing from Crisis to Sustainability*, New Haven, CT: Yale University Press.
SPREAD (Sustainable Lifestyles 2050 consortium) (2011), *Today's Facts & Tomorrow's Trends*, Wuppertal: European Commission's 7th Framework Programme: Socio-economic Sciences and Humanities.
Spycycles consortium (2009), *Cycling on the Rise: Public Bicycles and Other European Experiences*, Rome: Intelligent Energy Europe.
Thaler, R.H. and C.R. Sunstein (2008), *Nudge: Improving Decisions About Health, Wealth, and Happiness*, New Haven, CT: Yale University Press.
Tukker, A., G. Huppes et al. (2006), *Environmental Impact of Products (EIPRO): Analysis of the Life Cycle Environmental Impacts Related to the Final Consumption of the EU-25*, Seville: JRC /IPTS / and ESTO, p. 139.
Tuominen, P. and K. Klobut (2009), Country Specific Factors – Report of Findings in WP3, Deliverable 3.1 of the IDEAL EPBD project', Supported by Intelligent Energy Europe, available at http://ideal-epbd.eu/download/country_specific_factors.pdf (accessed 6 September, 2012).
UNEP (2001), Consumption Opportunities: Strategies for Change – A Report for

Decision-makers, Paris: United Nations Environmental Programme, Geneva: UNEP, p. 69.
UNEP (2009), *A Global Green New Deal*, Paris: United Nations Environment Programme, p. 16.
WBCSD (2008), *Sustainable Consumption Facts & Trends from a Business Perspective*, Geneva: World Business Council for Sustainable Development.
Weidema, B.P., M. Wesnæs, J. Hermansen, T. Kristensen and N. Halberg (2008), 'Environmental improvement potentials of meat and dairy products' in P. Eder and L. Delgado (eds), *Environmental Improvement Potentials of Meat and Dairy Products*, JRC/IPTS, available at http://www.saiplatform.org/uploads/Library/EnvironmentalImprovementsPotentialsofMeatandDairyProducts.pdf (accessed 6 September, 2012).
Wikipedia (2012), 'World Vegetarian Day' available at http://en.wikipedia.org/wiki/WorldVegetarian_Day (accessed 6 September, 2012).
World Bank (2012), 'Global commodity market outlook', *Global Economic Prospects*, January, Commodity Annex, Washington, D.C.
World Watch Institute (2010), *State of the World 2010. Transforming Cultures. From Consumerism to Sustainability*, Washington, DC: World Watch Institute.
WWF (World Wide Fund for Nature) (2008), *One Planet Mobility: A Journey towards a Sustainable Future*, London: WWF.
WWF, Ecofys, et al. (2011), 'The energy report. 100% renewable energy by 2050', report by WWF International, Ecofys, and The Office for Metropolitan Architecture.

6. Identifying relevance and strength of barriers to changes in energy behaviour among end consumers and households: the BarEnergy project

Martin van de Lindt, Sophie Emmert and Helma Luiten

INTRODUCTION

Energy is an expensive and increasingly scarce resource. Despite widespread recognition of diminishing oil reserves, a significant proportion of energy consumption happens through wasteful and inefficient practices. Public authorities, particularly the European Union (EU), have a key role to play in raising awareness of the issue and increasing energy efficiency for the sake of the environment, the economy and our health (European Commission, 2005). Consumers and households can also play an important role in reducing energy use by making changes in their energy behaviour.

This forms the background for the BarEnergy project. BarEnergy – an acronym for Barriers to changes in energy behaviour among end consumers and households – was an EU-funded collaborative research project, undertaken between 2008 and 2010. The project team comprised eight different organizations from six European countries (Table 6.1).

The project objectives were threefold. In the first place BarEnergy aimed to explore the strength and relevance of various barriers to change in consumer energy behaviour. An exploration of these barriers, together with relevant windows of opportunity, formed the basis for policy recommendations to overcome these barriers. Last but not least, the project wanted to contribute to methodological development in this area of research.

This chapter gives a bird's eye view of the project and summarizes the main findings; focusing on the methodology and the identification of various barriers to changes in energy behaviour among end users. First, we consider the relevance of the project, followed by the used starting points and concepts. The main results and recommendations are then addressed. We conclude with some reflections on the methodology.

Table 6.1 Project team of the BarEnergy project

Organization	Country
National Institute for Consumer Research	Norway
University of Surrey, Centre for Environmental Strategy	UK
The Netherlands Organisations for Applied Scientific Research	The Netherlands
Electricité de France	France
University of St Gallen	Switzerland
Central European University	Hungary
University of Groningen	The Netherlands
Centre for Sustainable Energy	UK

RELEVANCE OF THE PROJECT

Despite broad consensus around the fact that fossil energy resources are becoming scarcer, and that the use of these resources have very negative environmental effects, trends like the growing demand for electronic products and mobility are likely to drive the use of these resources upwards. Numerous efforts and initiatives to change this are undertaken, but for a number of reasons their success is often limited in time, place and/or magnitude.

One of the reasons is that top-down initiatives are sometimes successful (for example, labelling), but more often face implementation problems. These problems, which are partly intertwined, arise in many cases from lack of acceptance in the production chain, lack of consensus about the proven effects, lack of knowledge and trust among producers and consumers and so on. Bottom-up initiatives are sometimes very promising. For example, Passive Houses and community-owned energy companies have gained greater influence in recent times. However, scaling up in terms of bringing about a real change often fails. There are many reasons for this, but a very dominant one is the fact that the up-scaling can be threatening for the existing 'regime'. When such initiatives should scale up, this will have consequences for the markets at hand, the way companies have organized themselves and relations in the production, consumption chain and so on. Therefore, incumbents often feel more comfortable in the existing situation and are not willing to change to a great degree (Grin et al., 2010; Lindt and Elkhuizen, 2008; Tukker et al., 2008). A third reason for the limited success of initiatives is that changing energy behaviour is not only about removing technical barriers. On the contrary, derived from the already mentioned reasons, it can be argued that it is the result of a complex interplay between various levels (societal, structural, political, individual) and therefore between various actors and institutional arrangements.

The last point above has particular relevance for the BarEnergy study. Barriers faced by consumers in changing their energy behaviour go beyond the technological (such as the availability of energy-efficient alternatives); they can also be institutional (for example, low prices for unsustainable energy sources) and social (for example, a negative image that is related to a specific energy-efficient alternative). In the context of UK energy policy, Jackson (2006, p.89) states that:

> One of the biggest risks for government in relation to its energy policy targets is that of failing to understand the relationship between technological change, institutional change and social change. Energy systems cannot be thought of purely as technological systems. Energy technologies are embedded in institutional and social systems. Change in one implies change in the others. At the highest level, therefore, the role of the social sciences in energy research can be understood in terms of understanding and managing systemic change.

The BarEnergy research project therefore explores the barriers to consumers in changing their energy behaviour, and the interplay between the institutional and individual level. Households can reduce their energy consumption by saving energy, buying efficient appliances and shifting to more sustainable energy carriers. These distinct dimensions of behaviour face distinct barriers. In order for policy makers to be able to 'reach' the consumer more effectively we need to know more about the distinction between these barriers to change.

METHODOLOGY

From the issues raised in the above section and to meet the aims and objectives of the study, three starting points were defined.

Consumption Areas

The first starting point is the concentration on three consumption areas, identified by the European Commission for specific energy policy measures: domestic energy use, household appliances and personal transportation (limiting the fuel consumption of vehicles). These areas are responsible for a substantial part of the total energy consumption in the EU and within these areas contradictory trends and developments are to be seen.

Domestic energy use. The built environment accounts for more than 40 per cent of total energy use in the EU (based on 2002 figures). While, for example, Passive Houses[1] are gaining popularity in the Northern European countries, the Southern part of Europe is experiencing the reverse, with a move away from its traditional passive architecture aiming at a natural way for heating and cooling. This contributes to one of the most significant trends regarding domestic energy use in the EU: the significant increase in room air conditioning in the south of Europe.

Household appliances. In 1992 the EU made energy labelling mandatory for most household appliances. This has generally been regarded as an efficient way of increasing awareness of the environmental impact of consumption and the need to reduce resource use; and to promote the purchase of energy-efficient appliances. However, for energy labelling to be effective in all European markets, consumers must have an understanding of and confidence in the labelling scheme. It needs to be considered relevant and impartial, with a transparent approach from producers in meeting the labelling criteria. A further dimension in relation to household appliances and energy reduction is the potential to limit use (for example, turn off lighting, shower less).

Personal transportation (limiting fuel consumption). Road transport accounts for the second largest proportion of all energy consumption in the EU (based on 2002 figures). Carbon dioxide (CO_2) emissions from this sector grew by 20 per cent between 1990 and 2000. In terms of transport growth, the European Commission has calculated that without a change of policy and a yearly economic growth of between 1.2 per cent and 2.2 per cent between 2005 and 2050, personal transport will grow by 51 per cent and freight transport by 82 per cent (European Commission, 2009). Various fiscal and regulatory measures have been proposed and tested to curb this projected growth, for example, road pricing and reduced speed limits. Technological improvements in vehicle efficiency have also received much attention. However, the European Commission stated in 2011 that 'new vehicles have become more fuel efficient and hence emit less CO_2 per km than earlier models did in the past, but these gains have been eaten up by rising vehicle numbers, increasing traffic volumes, and in many cases better performance in terms of speed, safety and comfort' (European Commission, 2011, p. 12).

Strategies to Reduce Energy Use

The second starting point relates to the three main strategies to reduce energy uses, more or less according to the so-called 'Trias Energetica'. This is a simple and logical concept along three dimensions that helps to achieve energy savings, reduces dependence on fossil fuels and protects the environment. The first one is day-to-day behaviour to reduce energy in peoples' daily routines. Examples are turn down/switch off appliances and heating systems, take the bike instead of the car on short distances and so on.

The second dimension, energy-efficient purchasing, considers improvements in purchasing energy-efficient solutions. This includes consumers buying energy-efficient appliances and therefore relates to changes in purchasing behaviour. Because consumers have to 'invest' money to purchase the appliances, further opportunities can emerge concomitantly (Svane, 2002). For example, buying a house might provide an opportunity for upgrading the insulation of the property.

The third dimension is about moving towards often innovative sustainable and renewable energy technologies. In this dimension the consumers initially involved tend to be the 'frontrunners' in terms of environmental proactivity, sometimes called 'Dark Green Consumers' (Wüstenhagen, 2000). They often take a certain kind of risk because in many cases techniques and products are not mature and are suffering from 'growing pains'. In many cases the products are also not fully ready for the market and require considerable financial efforts. Examples are the first photovoltaic solar panels and the electric vehicles that are on the market now. The challenge is of course to scale these innovations up from niche to mainstream.

Individual and Institutional Approach

The third starting point refers to the combination of barrier types with the institutional approach. Based on the work of Crossley (1983), Shove (1998), Vine et al. (2003), the European Commission (2005) and Tukker et al. (2008), six major types of barriers influencing energy behaviour at the individual (household) level are distinguished, ranking from macro to micro perspectives.

Physical and structural barriers. Households are a part of, and their behaviour is dependent on, the general structure of society. The overwhelming majority are connected to electrical, telecommunications, water and wastewater networks. The degree of freedom and opportunities for individual households to act hinges largely on the physical environment. The availability of infrastructure and supply of products can promote particular behaviours and inhibit others. Political authorities and energy producers therefore have a responsibility in addressing these structural barriers to behaviour change.

Political barriers. Politicians create frameworks that underpin consumer behaviour. They issue laws, directives and develop regulations on national and European levels. These laws and regulations are then implemented by political authorities. Thus, political authorities determine the potential for change, and freedom for individuals to act.

Cultural-normative or social barriers. Energy-saving behaviour may not always comply with local customs and way of life. For example, Norway has the world's highest per capita energy use for lighting; a major contributing factor to this is the cultural aesthetics, which associates interior lighting and pools of light and shadow with a cosy atmosphere. Willingness to accept new technology may also differ across cultures, and energy-efficient appliances may not be perceived to increase one's social position.

Economic barriers. Some measures to reduce household energy use necessitate economic investment, and the willingness to pay represents a real barrier for some consumer groups. Investment in energy-efficient appliances may pay back over time, but this assumes an availability of funds and willingness to sacrifice these for a long-term gain. Furthermore, the 'green' option can sometimes be the more costly one – for example, in some countries energy from renewable sources has a higher unit price.

Knowledge-based barriers. To overcome barriers individuals need relevant information, or an understanding of how to obtain such information. The level of knowledge required to ensure uptake of measures and change in behaviour will vary. In some cases this may necessitate a clear grasp of the relevance of different actions and potential benefits.

Individual/psychological barriers. We all have our own limits as to what we would do to achieve a goal like saving energy. Such limitations stem from individual taboos, rooted in personal experiences or upbringing. Furthermore, energy-saving behaviour and actions may have negative connotations, being associated with discomfort or a restricted lifestyle. As such, the way information is portrayed – and subsequently understood or misinterpreted – has a fundamental link to this barrier.

The institutional approach is portrayed by the triangular model in Figure 6.1. Here society's institutional order is understood to be rooted in three major spheres: the state, the market and civil society. The triangle is a common analytical tool in sociology and related academic disciplines, and is applied in a variety of theoretical and empirical studies. For the purpose of the BarEnergy project domestic energy consumption sits at the heart of the triangle, being influenced by all three spheres. The six barriers to behaviour change can be placed within Figure 6.1. The physical and structural barriers incorporate elements of both the market and the state, while the political, cultural and economic barriers are associated with state, civil society and market, respectively. The two remaining barriers – knowledge and psychological – relate to the individual, rather than the institutional, level.

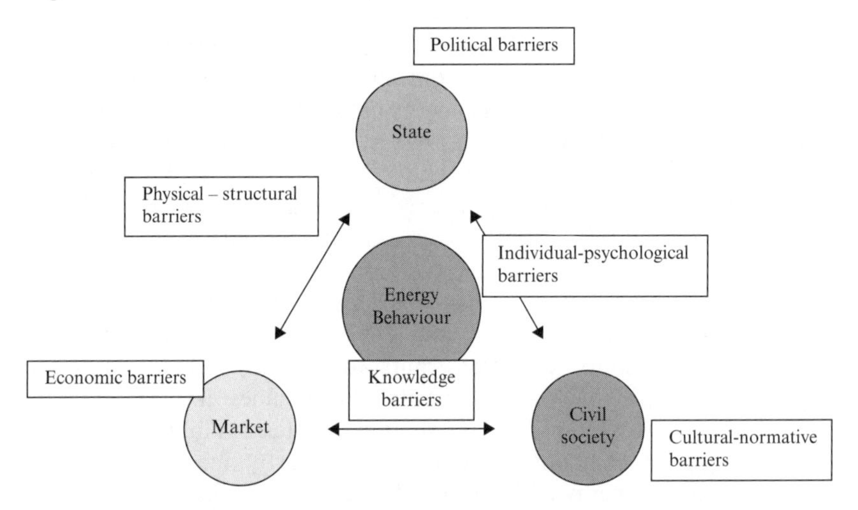

Figure 6.1 Complex interplay between behaviour, barriers, institutions and regimes

WORKING METHOD

Since BarEnergy was a transnational project, it was important to explore the different energy regimes in the participating countries. It makes a lot of difference if a country, for example, is an importer or an exporter of energy or if a country possesses renewable energy, which is of great influence on the energy mix: the proportion of fossil fuels, nuclear power, hydro power and other renewables. This kind of information provides an important contextual background for framing the research findings (Table 6.2).

Regarding the different levels of barriers in combination with the institutional approach, there was no one research method to cover all of this. Therefore, a so-called Empirical Triangulation was developed: a combination of three well-known research methods:

Semi-structured qualitative stakeholder interviews. The main objective of the qualitative interviews was to explore stakeholders' opinions about the relevance and strengths of barriers, and potential strategies to overcome these from the institutional perspective. The interviews also provided an opportunity to identify and verify the proposed barriers, and explore any conflicts of interest between different stakeholders. In the participating countries 162 qualitative interviews were held with strategic stakeholders from the scientific and technical community, energy suppliers and the construction industry, political authorities or semi-public institutions, planning departments and consumer and environmental organizations non-governmental organizations (NGOs).

Quantitative consumer survey. The aim of the questionnaire was to assess the strength and relevance of the barriers to change at the individual household/consumer level in each country. It also explored the extent of opportunity for change in the three core areas of study: reducing consumption (turn down and switch off), increasing energy efficiency and moving towards renewable energy. The representative quantitative surveys were conducted with a total of 15414 consumers across the participating countries.

Focus groups. The focus groups were aimed at identifying barriers and chances for reducing energy behaviour within various consumer groups. In total, 24 targeted focus group sessions were held in the participating countries. The groups were structured to include individuals with different 'windows of opportunity' and different social backgrounds. Windows of opportunity were defined according to whether the household was considered 'stable' or 'transitional', whereby the latter, for example, includes those in the process of undertaking building or refurbishment work, or moving home, therefore offering a potential increased opportunity for changing behaviour. Social background was based on social patterns and lifestyles as well on gender and age.

Figure 6.2 illustrates the research process and working method in detail, showing the link between the three empirical approaches, the three distinct energy behaviour types and the three domains in which the research was carried out.

Table 6.2 Energy regime summary

	Norway	UK	Netherlands	France	Switzerland	Hungary
Net import or export of electricity (2005)	Export (2005)	Import (2004)	Import (2002)	Export	Import	Export (2000)
Energy carrier for heating	Electricity Heating oil Wood	Gas	Gas District heating	Electricity Wood	Heating oil Gas Electricity	District heating Heating oil Coal
Energy mix (for electricity)	Renewable (99%)	Gas Coal Nuclear Renewables (3.1%) Oil	Gas Coal Nuclear Waste and renewables (1.4%)	Nuclear Renewables (14%) Oil and gas	Renewables (61%) Nuclear	Nuclear Coal Gas Renewables
Renewable sources	Hydro	Biomass Hydro Solar Wind	Biomass Wind Solar	Hydro Biomass Geothermal	Hydro	Solid biomass Solar Thermal
Regulated or deregulated	Deregulated	Deregulated	Deregulated	Deregulated	Regulated	Partly regulated

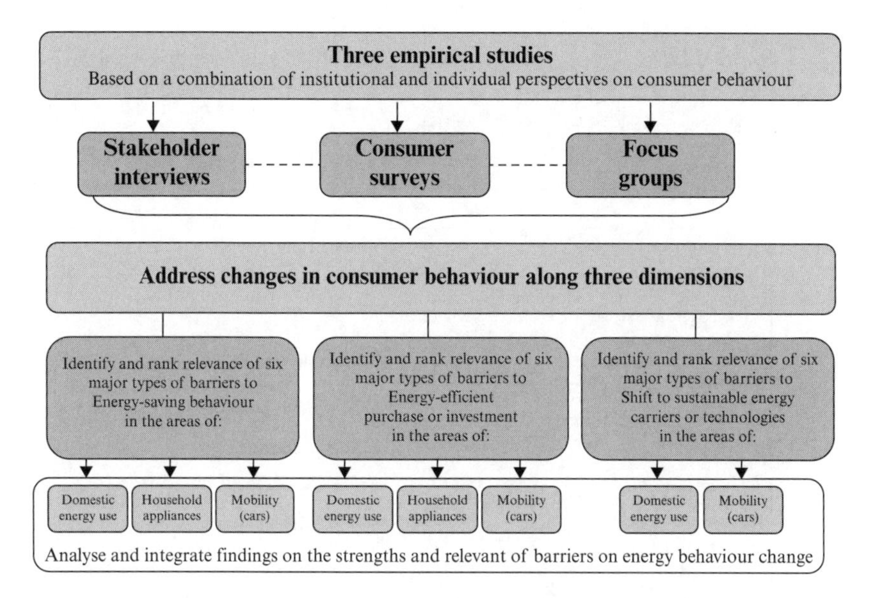

Figure 6.2 Research process and working method of BarEnergy

IDENTIFICATION OF STRENGTH AND RELEVANCE OF THE VARIOUS BARRIERS

At first glance the results from the main empirical research on the strength and relevance of the various barriers to the three strategies (saving, efficiency and change) suggest there is little difference between them. The same sets of barriers were repeatedly identified for all three strategies, as shown in Figure 6.3.

However, after a closer look at the empirical results, we learned in what way the meaning of the apparently equal barriers within the three strategies substantially differs from each other.

ENERGY SAVING: DAY-TO-DAY BEHAVIOUR

Day-to-day energy-saving behaviour is something every consumer could practise, at any time and with little effort, by making small adjustments to daily habits. It is the collective impact of these small, individual changes that offers significant potential in the effort to reduce energy consumption. This section identifies the main barriers that prevent individuals from contributing to energy saving.

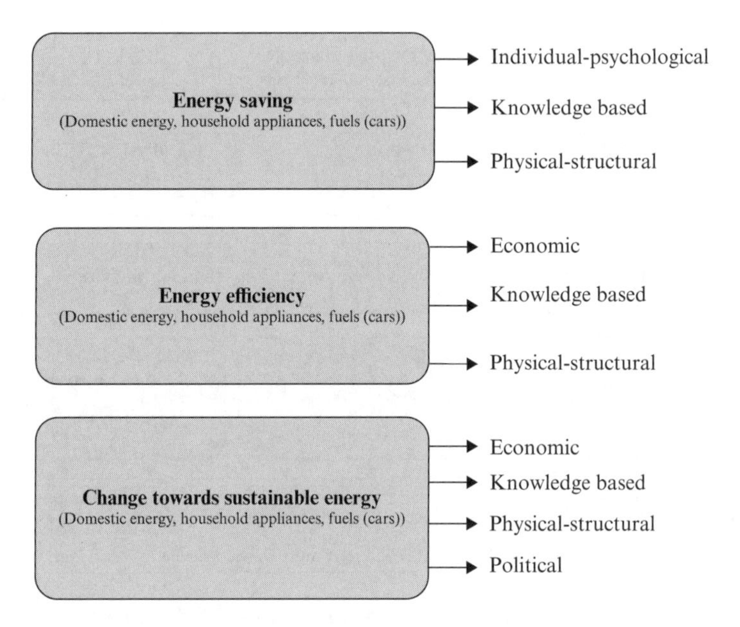

Figure 6.3 Three most important barriers for the three strategies

The analysis of the barriers to reduce consumers day-to-day energy behaviour from the stakeholder interview and focus group perspectives has shown that:

- In the area of domestic energy use most of the barriers with regard to heating behaviour are individual/psychological, cultural-normative and social, and knowledge based.
- In the area of household appliances individual, knowledge-based and physical and structural barriers are ranked highest when it comes to 'cooking and baking' activities and the use of electronic appliances.
- the area of personal transportation the main barriers are knowledge based, individual, political and physical-structural.

Domestic Energy Use

A key barrier identified through the consumer focus groups (in all participating countries) in the context of heating relates to the primacy of cultural norms for 'warmth', 'comfort' and 'wellbeing'. Based on the findings from all the empirical research we can deduce that personal comfort often takes higher priority than saving energy or environmental concerns. Another individual barrier that was identified is the personal effort that is required to save energy and ensuring this is a part of daily life and habits; for example, remembering to switch off lights and appliances, turning the heating down/off in rooms that are not in use. Focus group participants also noted a lack of evidence of energy-saving behaviour on a larger scale and wider, structural issues, for example, large public office buildings with the lights on 24/7 as having an impact on personal behaviour. Seeing energy being wasted on a large scale in their surroundings prompted some of focus group participants to question the value and meaning of their own personal efforts to reduce their consumption. Many argued that their efforts will make little difference and have little value in the scheme of things. Lack of knowledge amongst consumers about energy-efficient airing and ventilation was also identified as a barrier. Residents often lack knowledge about their dwelling and how to heat and ventilate it efficiently, resulting in wasteful heating practices. There was a widespread and strong desire amongst participants in the focus group discussions for knowledge about what improvements could be made to their own home, and the potential benefits. Here participants tended to focus on physical and technological improvements that could be done to their homes rather than on changing behaviour and habits. There seemed to be a distinct lack of knowledge about personal consumption patterns, and a technical solution seemed preferable to a change in behaviour.

Household Appliances

The subject of cooking behaviour was chosen as an area of study based on a survey by Bush et al. (2007). The survey explored the potential energy savings related to consumer behaviour in using appliances and performing a range of household tasks. It showed that the greatest potential for savings (50 per cent) were in lighting and cooking/baking. Based on the empirical results, the main barrier to saving energy in cooking and baking behaviour appeared to be the lack of will (Switzerland, Hungary). Interviewees suggested that consumers often lack a clear and informed idea of how they can influence energy consumption as individuals. This also applies to cooking and baking. However, representatives from NGOs and semi-public organizations maintained that information is available if the consumer actively seeks it. Again, it therefore seems that both a communication and a motivational challenge exists.

Regarding the 'use of appliances', the data showed that switching off appliances was a strongly habitual issue. The consumer survey explored the concept of habitual practices as a barrier to energy-saving behaviour.

The survey results also suggest that individuals feel a relatively strong moral obligation to turn lights off (personal norm) and that it is expected of them (subjective norm). The subjective and personal norms to turn off lights were relatively strong compared to other kinds of behaviours. Analysis of the survey results shows that psychological factors explain 49 per cent of all the variance in lighting behaviour. This suggests that the psychological barriers are very pertinent when it comes to switching off lights, with 'habit' being the single most influential factor. The findings from the stakeholder interviews support this result, with individual/psychological barriers being ranked most important in the context of switching off appliances more generally.

Personal Transport

Under the topic of short distance driving, and choosing public transport instead of the car, the individual/psychological barriers came through strongly in the data. Convenience factors associated with time, hassle, planning and flexibility contributed to a general preference for the private car. This tied in with cultural barriers related to car use, reflecting important traits of modern lifestyles. There were also political barriers and it was argued that city planning was often too focused on the private car. This leads to the general perception that the infrastructure is not sufficient to support alternative means of transport, even in urban areas with some level of public transport services. When it comes to carpooling and car sharing, the main barriers identified were also individual/psychological in nature (related to the pre-planning needed, inflexibility, lifestyle issues, safety concerns). Knowledge (a lack of information about the services) and physical structural (poor quality of networks) barriers were also evident. As with the other two topics under transportation, it could be argued that the service must first be improved for it to become a viable alternative to the use of private cars, thus addressing both the physical-structural and individual/psychological barriers. Where sufficient car-sharing and carpooling networks already exist, a lack of awareness can prevent them from succeeding.

Closing

Combining the findings from this empirical research suggests individual/psychological barriers are the most prominent across all three fields (domestic energy use, household appliances and private transportation). Personal comfort and convenience often appear top priority for consumers, ahead of environmental, and to some extent, economic concerns. However, it could be argued from the findings that the macro-related barriers are of higher importance as these influence and shape the choices of the individual consumer, thus framing the barriers at the lower level.

With regard to day-to-day energy-saving behaviour, consumers need to be convinced that the collective impact of small adjustments in individual practices could offer significant potential in the effort to reduce energy consumption. The knowledge level on energy behaviour within social networks and the role model of peer groups in civil society have a key role to play in changing personal habits and cultural norms. Based on the institutional approach, portrayed by the triangular model, Figure 6.4 illustrates that the key barriers to day-to-day energy-saving behaviour mainly reflect at the level of civil society.

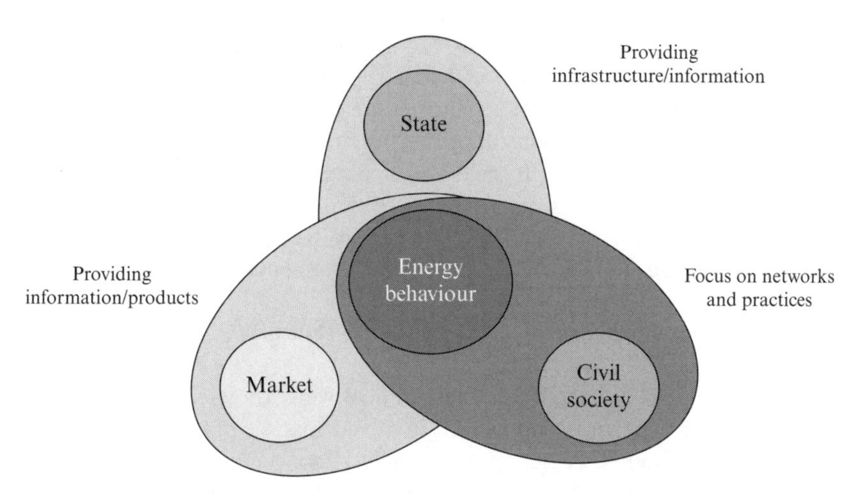

Figure 6.4 Main policy directions towards energy saving

However, initiatives at consumer level must be combined with market support by offering information and solutions that create opportunities for energy-saving behaviour. The state, therefore, also has a responsibility in providing the required resources and infrastructure to effect this change.

ENERGY-EFFICIENT PURCHASING DECISIONS

Different from the day-to-day behaviour is the behaviour at moments of purchasing. Energy-efficient technology already exists to help reduce energy consumption in households. By maximizing technology potential with more widespread behaviour changes in practices and habits, the residential sector could generate huge energy savings compared to other sectors (UNECE, 2009). Therefore, we researched the barriers to purchasing energy-efficient solutions. In the area of domestic households the project looked at energy-efficient refurbishment. In the area of household appliances the purchase of energy-efficient appliances was researched. In the area of personal transportation we researched the purchase of average fuel-efficient cars (the hybrid cars are the subject of the next section 'Change to sustainable energy sources'). The analysis of the barriers to purchasing energy-efficient solutions from the stakeholder interview and focus group perspectives has shown that:

- Most of the barriers with regard to energy-efficient refurbishment decisions are knowledge based, economic, physical and structural.
- For purchasing of energy-efficient appliances economic, knowledge-based and physical and structural barriers are ranked highest.
- The main barriers regarding the purchase of fuel-efficient cars are political, economic, physical and structural.

The quantitative survey specifically looked at the attitude and motives for purchasing energy-efficient light bulbs and green energy.

Domestic Household (Energy-efficient Refurbishment)

The predominant barriers that emerged here were knowledge based, such as the lack of expertise amongst all involved market players – including professionals, homeowners and installers. Economic barriers, such as the (perceived) high initial cost of measures, also play an important role. Consumers often face a barrier in financing home improvements, either because they do not have the capital available or cannot secure a loan. However, the perceived high costs can be misconceived. Apart from having difficulties concerning high initial costs, consumers do not often adopt a long-term perspective and take account of life cycle costs of measures. This can stem from a lack of knowledge or interest. Thus, the economic and individual/psychological barriers are closely interlinked. Physical and structural barriers were ranked as the third most important factor, like lengthy decision-making processes in joint ownership, complicated collective decision making and lack of skilled professionals. Based on the barriers identified, it is recommended that the market forms the basis for the policy framework, in particular the evident skills gap in the area of energy-efficient refurbishment needs to be addressed. This could be achieved through appropriate training and education systems for professionals and installers, such that they are equipped to provide consumers with correct and relevant information about energy-efficiency measures. At the same time, on the demand side consumer awareness needs addressing. This could be achieved through awareness-raising initiatives, combined with (free) energy audits. However, such initiatives need to go beyond the general to ensure tailored information is targeted at specific consumer groups and windows of opportunities, for example, buying a new house/moving home, transitions in family circumstances and so on. This necessitates joint collaboration between professionals, the state and regime players, taking advantage of popular media and maintaining a consistent and clear message.

Household Appliances (Purchase of Energy-efficient Appliances)

The growing demand for consumer electronics in recent years poses a real dilemma for governments as they try to reconcile lifestyle aspirations and concerns over environmental degradation and energy security. The (perceived) higher price of energy-efficient appliances emerged as the most important barrier. Besides the high initial cost, the low price of electricity was mentioned by the stakeholders and focus group participants as an important barrier. In Norway and Switzerland especially the low price of electricity was associated with limited incentive to invest in energy efficiency. Some stakeholders considered the price of electricity would have to be substantially higher than current levels in order to act as any incentive to behaviour change. Many stakeholders interviewed felt that salespeople were often not sufficiently trained to competently advise consumers on their purchasing decisions. Respondents from Norway, France and Hungary rate this structural barrier as important, whereas in the other countries it was rated of medium importance. Stakeholders (and some of the focus group participants) felt that it is the responsibility of the salespeople to point out the relevance of the labels and to answer questions regarding the energy consumption of appliances. Furthermore, levels of disposable income and lack of awareness of, or appreciation for, life cycle costs were also relevant in the context of price. While the energy label was acknowledged to be a success in communicating energy-efficiency ratings to the consumer, there was some level of confusion with the application of the expanded scale (like label A++). Consumer purchase decisions are influenced by a wide range of factors. The extent to which energy efficiency is a consideration appears to vary according to (amongst other things) whether the appliance is considered essential to everyday life (such as the washing machine) or a luxury product.

The quantitative survey explored purchasing behaviour on energy-efficient light bulbs and green energy. It shows that in Norwegian households the average percentage of households accepting energy-efficient light bulbs is lower than in other countries. In Norway the average percentage is 33 per cent, while in the other countries the average percentage is around 50 per cent. Participants indicate that they feel that it was important others might expect them to purchase energy-efficient light bulbs. This was felt strongest in the UK and Greece, while in Norway it was felt weakest, indicating that Norwegians felt a weaker expectation from others to purchase energy-efficient light bulbs than participants from other countries. Most participants felt that they were able to purchase and use energy-efficient light bulbs. Participants' problem awareness was moderate. It appeared that most participants did not think that the use of inefficient light bulbs caused serious environmental problems. However, most participants did seem convinced that they would be able to contribute to solving these problems by using energy-efficient light bulbs. Both problem awareness and outcome efficacy were lower in the Netherlands and Norway than in the other five countries. Participants did not perceive a strong moral obligation to use energy-efficient light bulbs. Norwegian participants in particular reported no strong sense of moral obligation to use energy-efficient light bulbs.

Concerning the purchase of green energy, more participants in the Netherlands and Norway used energy from sustainable sources than participants in other countries. Nearly 40 per cent of all participants used energy from sustainable sources. In the Netherlands and Norway this figure was close to 60 per cent. One of the demographic factors that appeared to influence sustainable energy use is income level. The use of energy from sustainable energy sources increases with income level in most countries. Hungary and the Netherlands were exceptions to this trend.

There is a strong role for the state and regime in addressing demand-side issues associated with the purchase of energy-efficient appliances.

Personal Transportation (Purchase of Fuel-efficient Cars)

Continuing increases in the volume and number of private vehicles pose a significant problem for governments across the EU. Overall, the main barrier to the purchase of energy-efficient cars is related to insufficient regulations on fuel-efficiency standards. As a result, inefficient cars are still affordable to consumers and, as consumers are mostly guided by a cost/gain perspective, there is little incentive to opt for a more efficient vehicle. Individual/psychological barriers are also important. In the stakeholder interviews fuel efficiency was identified as a low or non-existent consideration for car buyers. Horsepower and brand were considered far more important. This may be the result of advertising being focused on inefficient cars or the perceived low status symbol of energy-efficient cars. Based on these barriers, one of the main recommendations for policy makers focuses on the product – consumer chain is to introduce incentives for the purchase of fuel-efficient cars, and penalties for buying inefficient cars. The state/regime has a crucial role in implementing such incentives. Initiatives include: scrappage schemes; exemption from import and excise duties; free parking for fuel-efficient vehicles; exemption from road tolls; and permission to use public transport lanes. On the supply side retailers have an import role. It is recommended that the system of higher profits/ commission on larger and more expensive cars is re-evaluated.

Furthermore, sales personnel need to be well trained to give consumers correct and relevant information about rules, regulations, life cycle costs and so on. The manufacturing industry is already sending a positive message, with increasing emphasis being put on the manufacture of fuel-efficient cars and developing more appealing models. The government should support these kinds of developments and develop transparent policy on efficiency standards.

Closing

There are multiple examples of policy measures implemented in EU countries to promote energy efficiency. The results from the quantitative study showed that there is a clear difference in consumer perceptions of 'pull' and 'push' measures. Pull measures were generally deemed very acceptable, fair and effective, while push measures were considered the exact opposite.

Based upon the findings above from this empirical research, policy directions to address barriers for energy-efficient purchase decisions in and around the house should focus on the market, such that products have to meet specifications and retailers act as sources of information. At the same time, consumer demand for energy-efficient products has to be stimulated. Probably other relationships, other business models, between the market and the consumer should be developed in order to better meet personal attitudes in an energy-efficient way. Figure 6.5 illustrates that the key barriers within the purchasing process of energy-efficient appliances mainly reflect on the market level. There should also be greater focus on green innovation of product requirements, retailers becoming a source of information on energy efficiency and stimulating consumer demand on green purchases.

However, the state and civil society also play an important role. Policy measures must also therefore take account of social networks and

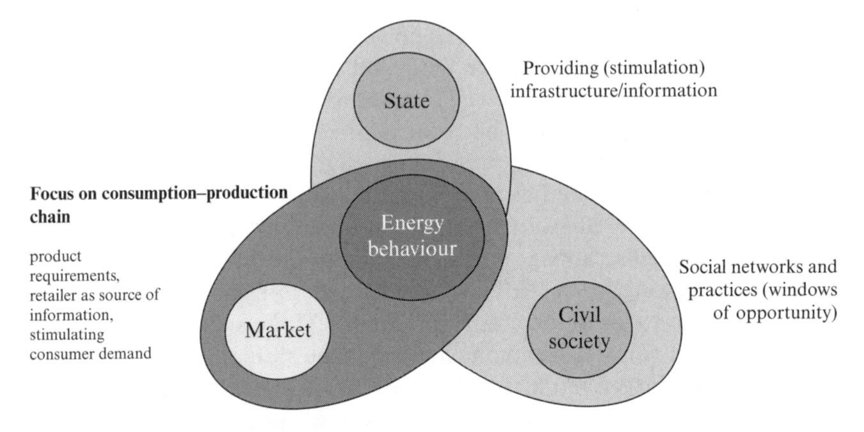

Figure 6.5 Main policy directions towards energy efficient purchasing

practices, and focus on maximizing potential 'windows of opportunity' (moving home, refurbishing and so on). In some of the focus groups participants seemed willing to embrace energy-efficient purchases at opportune moments. Reduced payback periods are crucial to ensuring this willingness is acted upon. This asks for a joint collaboration between professionals, the state and regime players, taking advantage of popular media and maintaining a consistent and clear message.

CHANGE TO SUSTAINABLE AND RENEWABLE ENERGY TECHNOLOGIES

The rising costs, scarcity of supply and negative environmental impacts of fossil fuels necessitate a shift to alternative energy sources. The domestic and road transport sector account for a significant proportion of total energy consumption. The project therefore explored the barriers to using sustainable energy in both the home (including the installation of photovoltaic panels, the purchase of green power and the construction of energy-efficient houses) and private transportation (focusing on hybrid and fuel-efficient cars).

The analysis of the barriers to change to sustainable energy from the stakeholder interviews and focus group perspectives showed that:

- Most of the barriers with regard to the installation of photovoltaic panels and the construction of energy-efficient houses are economic, political and knowledge based.

- The main barriers to the purchase of hybrid cars and green fuel cars are physical-structural, economic, political and knowledge based.

Domestic Energy Domain

Driven by advances in technology and increases in the scale and quality of manufacturing, the cost of photovoltaics has declined steadily since the first solar cells were marketed (Swanson, 2009). However, the installation of renewable energy technologies, such as photovoltaic panels, still entails significant upfront costs. The most important economic barrier mentioned in both the stakeholder interviews and focus group discussions was the high initial investment. The cost of the technology and its relative lifetime also mean that, compared to normal electricity, solar power has a very high unit price. An additional economic barrier identified is the lack of financial incentives offered to potential consumers, which could help to overcome the economic constraint. The most important political barrier to emerge was the low level or lack of government support and inconsistent political framework. This barrier was mentioned in all countries of study. Stakeholders also identified a lack of competency in sales and marketing. Supply-side offers are often poorly structured and not focused on customers' and consumers' needs. There also appears to be a distinct lack of professional installers. This barrier correlates strongly with the current low demand for the technology in a negative feedback loop: low demand results in poor availability/quality of supply, which subsequently further discourages demand.

Also with regard to the construction of low energy houses, both stakeholders and focus group participants argued, the main economic barrier to the construction of low energy homes is the high initial investment costs. Insufficient government support is the most important political barrier to building energy-efficient houses. Stakeholders felt that governments need to provide greater support to enable the more expensive energy-efficient construction methods. The consumer survey showed that a majority would approve of tougher political regulation to enforce energy-efficiency standards for new housing developments. Individuals considered such measures to be acceptable, fair and effective in reducing household energy use. In general the lack of knowledge of almost all stakeholders (owners, installers and architects) was considered an important barrier to building energy-efficient houses. Finding an architect who is familiar with low energy homes is difficult, and can take multiple attempts and more effort.

Green Fuel Cars

The European Commission has a target of a 20 per cent reduction in emissions in the EU by 2020. The most important barriers regarding a change towards green fuel cars exist at both the institutional and individual levels. At the institutional level a lack of infrastructure and appropriate facilities is a significant barrier preventing green fuels from becoming mainstream. The main barrier for hybrid cars is the higher price tag, followed by the fact that there are few models to choose from. From a market perspective there is a need to increase the supply of hybrid and electric vehicles, to expand both the new and second-hand market. The prevalence of economic barriers suggests that, from a policy perspective, financial incentives are needed to encourage consumers to shift to sustainable fuel cars.

On an individual level the lack of knowledge and trust in hybrid and electric vehicles, and scepticism surrounding the sustainability credentials of biofuels emerged as significant barriers. The potential threat of biofuels to food production is one example of this scepticism that was raised in the focus group discussions and interviews. This suggests a need for improvements in the quality and accessibility of information, at the consumer level, on technical, environmental and ethical issues related to the production of biofuels and driving electric and hybrid cars.

Closing

In order to facilitate the change to sustainable and renewable solutions policy measures must combine a top-down (state) and bottom-up approach (Figure 6.6). Focus should be on state governance ensuring opportunity for new initiatives, with a clear and consistent energy policy framework, and rewarding frontrunners at the consumer, business and research levels. This should be supported by using social networks to reach and engage consumers and targeting specific groups within a specific context to maximize windows of opportunities. The 'market' also has a role in ensuring retailers are equipped to act as trained consultants in advising consumers.

REFLECTIONS ON THE METHODOLOGY

One of the project objectives was to contribute to methodological development in this area of research. After the presentation of the main results and the main directions of recommendations in the paragraphs above, we conclude with some reflections on the methodology.

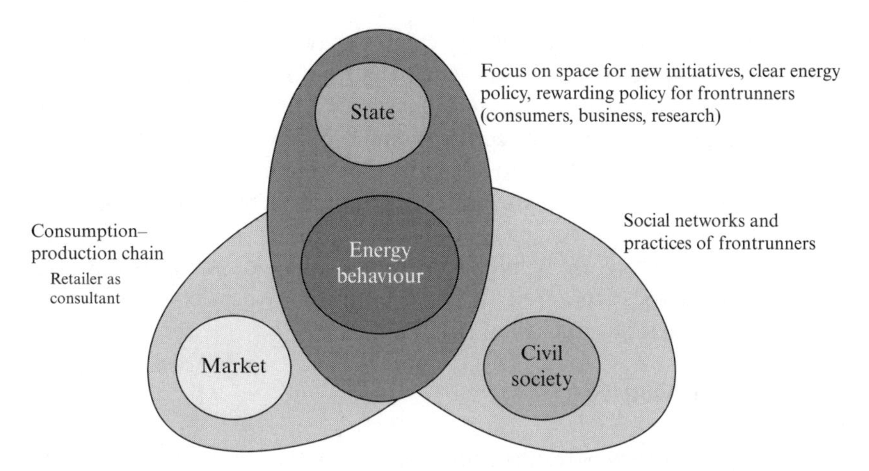

Figure 6.6 Main policy directions towards the change to sustainable and renewable solutions

The use of three different methods for collecting information is considered to be one of the key strengths of this project. The stakeholder interviews provided the expert perspective; the quantitative survey provided a representative view on drivers for change; and the focus groups gave in-depth information about consumer behaviour. These methods did not always provide the same results. In this section two contradictory results of the different studies are discussed. Some clarification on these contradictions is provided as well as a discussion on the added value of using multiple methods.

Knowledge

Participants in the focus groups stressed that they were willing to take steps to reduce their energy consumption, if only they knew how. This emphasis on knowledge as an important barrier is at odds with the results from the quantitative survey and with previous (quantitative) research. In the survey, for example, it was found that knowledge about solar panels was a very poor predictor of the acceptability of subsidies for the installation of the panels. Previous quantitative work also reports that knowledge levels have little predictive value for pro-environmental behaviours.

Clearly, consumers require a certain amount of knowledge to effectively reduce their energy consumption; but most participants in the focus groups appeared to have a basic grasp of this topic. The main barrier seemed to be not knowing exactly which behaviour is most appropriate and effective, rather than not knowing what to do at all. Individuals who do not know exactly how to reduce energy consumption are still able to make changes in the right direction, assuming they are motivated to do so. This may explain why knowledge levels offer little predictive power in quantitative models that include motivational factors. Less motivated individuals may use a lack of specific knowledge as justification for not reducing their energy use, while in reality their lack of action is down to a lack of willingness to act. Thus, lack of specific knowledge can be an excuse for not acting rather than a major barrier.

Environmental Motives

The quantitative survey results show that consumers are motivated by the environmental consequences of their behaviour. When asked directly, participants indicated that environmental impacts are an important reason for them to reduce their energy consumption. Furthermore, the extent to which respondents regarded certain behaviours as environmentally damaging was highly predictive of whether they engaged in these behaviours (including reductions in car use and the use of energy from sustainable sources).

However, the focus groups provided different results. Focus group participants emphasized personal benefits (money, comfort) rather than environmental benefits as the main reason for changing behaviour. Moreover, in several countries the outcome of the focus group discussion was that environmental consequences in themselves are not sufficiently motivating to trigger behaviour changes.

One explanation for this difference may be the fact that during the focus groups participants were asked to compare the different motives, while in the quantitative survey respondents only had to indicate to what extent environmental concerns were important to their decision making. Therefore, the participants in the focus groups may have overemphasized personal benefits simply because they found them more important than environmental consequences; not because environmental consequences were unimportant.

Categorization of Barriers

The categorization of the barriers is interesting and relevant, but the results show that these are interconnected. This explains in part why the 'solutions' are not simple to identify. For example, political and structural barriers overlap, where stakeholders perceived government to have an important role in shifting towards more sustainable infrastructures. Further research on this topic is necessary to provide better insight into the barriers and their interdependence with opportunities. It may lead to new definitions of barriers and opportunities, as well as the development of energy policy.

It is also recommended that the research should not be limited to the housing/domestic sector. The commercial sector also has high energy consumption. It consists of a large variety of buildings, belonging to various economic sectors. Each has special characteristics and requirements with respect to the physical structure and ownership, management and so on. For a sound and consistent policy towards energy saving, efficiency and behaviour change, further research to barriers, opportunities and motives is absolutely necessary.

NOTE

1. The Passive House is a voluntary building standard developed in Germany in the early 1990s by Professor Bo Adamson (Sweden) and Professor Wolfgang Feist (Germany). Built according to this standard, the need for heating will be reduced by 90 per cent together with the preservation of sufficient indoor air quality conditions. The remaining heating and hot water demand can be covered completely by renewable energies.

REFERENCES

Bush, E., B. Josephy and J. Nipkow (2007), *Energetisches Einsparpotenzial von főrdermassnahmen fűr energieeffizierte Haushaltgetd́e*, Felsberg and Zurich: Zutichi Energie GmbH ane ARENA.

Crossley, D.J. (1983), 'Identifying barriers to the success of consumer energy conservation policies', *Energy*, **8** (7), 533–46.

European Commission (2005), 'Doing more with less', Green Paper on energy efficiency, European Commission, Brussels.

European Commission (2009), *Towards an Integrated, Technology-led and user-friendly Transport System*, Directorate-General for Energy and Transport, European Commission, Brussels.

European Commission (2011), 'Impact Assesssment: roadmap to a single European transport area – towards a competitive and resource efficient transport system', European Commission staff working paper, Brussels.

Grin, J., J. Rotmans and J. Schot (2010), *Transitions to Sustainable Development: New Directions in the Study of Long Term Transformative Change*, New York and Oxford: Taylor & Francis, and Routledge, (with contributions of D. Loorbach and F. Geels.)

Jackson, T. (2006), *An Agenda for Social Science Research in Energy*, Summary of a Research Council Workshop, 6 April, Engineering and Physical Science Research Council's energy programme.

Lindt, M.C. and B. van de Elkhuizen (2008), *Barriers and Chances for Energy Reduction in the Building Sector*, Delft: TNO (in Dutch).
Shove, E. (1998), 'Gaps, barriers and conceptual chasms: theories of technology transfer and energy in buildings', *Energy Policy*, **26** (15) 1105–112.
Svane, Ö. (2002), *Nordic Households and Sustainable Housing – Mapping Situations of Opportunity*, TemaNord Report 2002: 523, Copenhagen: Nordiska Ministerrådet.
Swanson, R. (2009), 'Photovoltaics power up', *Science*, **324**, 15 May, 891.
Tukker, A., M. Charter, C. Vezzoli; E. Stő and M. Anderson (eds), (2008), *System Innovation for Sustainability: Perspectives on Radical Changes to Sustainable Consumption and Production*, Sheffield: Greenleaf Publishing.
UNECE (2009), *Green Homes – Towards Energy-efficient Housing in the United Nations Economic Commission for Europe region*, Geneva.
Vine, E., J. Hamrin, N. Eyre, D. Crossley, M. Maloney and G. Watt (2003), 'Public policy of energy efficiency and load management in changing electricity business', *Energy Policy*, **31** (5), 405–30.
Wüstenhagen, R. (2000), *Ökostrom von der Nische zum Massenmarkt*, Zurich: Vdf Hochschulverlag AG.

7. Collaborative (and sustainable) behaviours: grassroots innovation, social change and enabling strategies

Ezio Manzini

INTRODUCTION

In 1973 Ivan Illich organized a seminar in the Centre for Intercultural Documentation (CIDOC), which he had founded in Cuernavaca in Mexico. In his book *Tools for Conviviality*, which grew out of this event, he proposes an original vision of technology and the tools it offers, or rather, that technology could offer if it were more intelligent than it is currently: 'Give people tools that guarantee their right to work with independent efficiency' and that 'allow the user to express his meaning in action' (Illich, 1973, p. 35). This idea, which focuses on technology as a tool to give people independence in both making things and creating meanings, is the basis of Illich's proposal for a convivial society. That is, a society that would be 'the result of social arrangements that guarantee for each member the most ample and free access to the tools of the community and limit this freedom only in favor of another member's freedom' (Illich, 1973, p. 25).

There are several reasons why, today, Illich's reflections should be rediscovered and discussed in light of the way in which society has changed and is presently changing. To this end this chapter considers emerging collaborative (sustainable) behaviours, and the enabling strategies capable of supporting them towards the achievement of mainstream status, as opposed to their current status as predominantly active minorities.

ACTIVE MINORITIES

In its complexity and idiosyncrasy, the whole of contemporary society can be seen as a huge 'laboratory of ideas' for everyday life: ways of being and doing that express a capacity to formulate new questions and to find new answers. Among such cases there are some where this diffuse creativity has found a way of converging in collaborative activities: ways of living together where, in order to improve wellbeing, spaces and services are shared (as in the examples of collaborative housing); production activities are based on local capabilities and resources, but linked to wider global networks (as occurs with certain typical local products); a variety of initiatives operate focused on healthy, natural food (from the international Slow Food movement to the spreading of a new generation of farmers' markets and community-supported agriculture programmes); self-managed services operate, for example, childcare services (such as micro nurseries: small playgroups run by the parents themselves) and services for the elderly (such as home sharing where the young and the elderly live together); new forms of socialization and exchange (such as Local Exchange Trading Systems – 'Lets' and Time Banks); alternative transport systems to the individual car culture (from car sharing and carpooling to a rediscovery of bicycle potential); networks linking producers and consumers both directly and ethically (like the worldwide fair trade activities). The list could continue, touching on every area of daily life and emerging all over the world (to read more about them see Desis, 2012).

In the past decade this phenomenon has been observed and analysed by several research studies. Of particular relevance in this regard, for this chapter (and its author), are two large research programmes: EMUDE[1]and CCSL[2] as well as the design and research activities of an international network of design laboratories: the DESIS Network[3].

A clear message to emerge from these research programmes is the notion that there exists a dynamic new form of creativity: a diffused creativity put cooperatively into action by 'non-specialized' people, which takes shape as a significant expression of contemporary society, which might collectively be termed creative communities. That is, groups of innovative citizens organizing themselves to solve a problem or to open a new possibility, and doing so as a positive step in the social learning process towards social and environmental sustainability (Meroni, 2007). These creative communities are a component of a broader wave of changes that is internationally defined by the expression 'social innovation': ideas that work in solving socially relevant problems, and that do so in socially relevant ways (Mulgan, 2006).

CREATIVE COMMUNITIES

Creative communities are initiatives where, in different ways and for different motivations, some people have reoriented their behaviours and their expectations in a direction that appears as a positive step towards sustainability, with the unprecedented capacity to bring individual interests into line with social and environmental ones. In fact, even if each one of them is driven by the quest for solving very concrete problems, all of them present a fundamental side-effect: they reinforce the social fabric and, more generally, they generate and put into practice new, and more sustainable, ideas of well-being: a wellbeing coherent with major guidelines for environmental sustainability, including positive attitudes towards sharing spaces and goods; a preference for biological, regional and seasonal food; a tendency towards the regeneration of local networks and the creation of highly resilient distributed systems (Vezzoli and Manzini, 2008). For all these reasons, the promotion of creative communities – and promising initiatives they generate – can be considered as a viable strategy for sustainable development.

During the last 5–10 years these promising initiatives spread, reinforcing a shift from active minorities (signals of change) towards becoming vital and recognizable solutions to present and urgent problems, such as social integration, urban regeneration, health care, food accessibility, renewable energies and water management. At the same time they have generated broader visions of possible sustainable futures: a new scenario shaped and driven by this kind of social innovation that has the defining features of being small, local, open and connected (see the SLOC Scenario in Manzini, 2010, that is, 'Small Local, Open and Connected').

This evolutionary trend, from weak to broad and strong signals has been largely driven by two very diverse events; (1) the increase of connectivity and, (2) the diffusion of networks throughout society, on the one hand, and the growing pressure of the economic, social, environmental crisis, on the other.

HYBRID SOCIAL NETWORKS

The past decade has witnessed the explosion of social networks. But, in the last few years, their quantitative growth has been paralleled by an equally relevant qualitative evolution, characterized by a progressive shift from the digital sphere, where they were born (that is, in the domain of information generally, such as the flow of music, texts, photos, videos, avatars and so on) to a hybrid, digital/physical sphere where digital services and platforms connect people in the virtual sphere and enable them to meet and collaborate in the physical one. This evolution continues to generate a new large wave of innovation where social and technological components are tightly interwoven.

For a long period understanding and practice relating to creative communities and social networks remained separated. But over the last five years the distance between the two has started to shrink; with creative communities and social networks converging to the point where now they largely overlap. The result is a hybrid socio-technical innovation that continues to generate a growing number of practical solutions and general visions (Bauwens, 2006, 2007).

The first result of this evolution is that the traditional gap between mainly low-tech 'traditional' community-based innovation and the high-tech initiatives (based around social networks and peer-to-peer (p2p) organizations) has closed with the existence of overlaps increasing substantially. The more that this and more community-based innovations flourish in specifically developed digital environments, the more digital environments are developed capable of supporting people to collaborate and exert positive impacts on society, the economy and the environment (Baek, 2010).

If we consider today's multi-layer crisis (that is, where the economic and financial problems collide with and exacerbate environmental and social challenges), we can observe that these cases of social innovations are indicating promising and viable directions. Of course, we don't know what will happen, which directions people will search and where they will arrive. Nevertheless, we can observe that the worldwide movements that characterized 2011 (for example, Arab Spring, Indignados in Europe, Occupy Wall Street (OWS) in the US – Castells, 2012; Joffé, 2013 have been, and are, based on social problems and a new technologies mix that is very similar to the one generated by the everyday life innovation that we are discussing here. The leaders of these movements can also be seen as social innovators who invented brand new modalities of political expression (and therefore innovated politics). Finally, it can be noted and underlined that several OWS movement participants explicitly referred to the economy of creative communities and social networks as examples of the new economy they sought to establish.

Although the discussion of these political events lies outside the scope of this chapter, in order to complete the background on which my discussion will be developed, they must nevertheless be considered. Whatever their direct political impact will be in the future it is clear that they have had and, at the time of writing, continue to have a strong impact on ideas and mentalities. That is, they create the precondition of even larger social, economic and political changes.

IDEAS OF WELLBEING

As we have seen, the new forms of organizations appearing from the merge between creative communities and social networks can be considered in different ways from different viewpoints. From a 'behavioural' point of view it is evident that all of the organizations are characterized by the main role played by active and collaborative participants. That is, people who participate in an active and collaborative way to the definition of the goals to be achieved by the organization and who bring their contribution to the realization and delivery processes (Boyle and Harris, 2009; Boyle, et al., 2010). Clearly these emerging behaviours are very different from the (still) dominant ones, especially those behaviours relating to the mainstream 'modern' idea of wellbeing. This point about the Western traditional idea of wellbeing (which from now on will be referred to as 'modern wellbeing') is crucial and is explored further during the course of this chapter.

First of all, it is clear that modern wellbeing, although widely recognized as socially and environmentally unsustainable, is nonetheless still very powerful. The most obvious and discussed motivation for this is that modern wellbeing, and the expectations it generates, is supported by powerful economic forces. But a second, more profound motivation has to be considered too: modern wellbeing has been, and in many ways still is, very successful in practical terms. In fact, it drove a transformation that, for many people, has meant a concrete, tangible improvement in their quality of life.

This modern wellbeing background can be summarized in four steps: (1) human beings are individuals endowed with needs and wishes, (2) technology and economy should provide solutions capable of fulfilling those needs and satisfying those wishes, (3) producing better and cheaper products, and spreading them in society, enables the realization of (or, at least, an approximation to) a 'democracy of consumption' based on opportunities for everyone to live better, where – in this conceptual and practical framework – to live better means (4) to use and consume more products and services. In short, the power of modern wellbeing is that it is proposed as a kind of product: a wellbeing that can be served and individually consumed.

In this framework the role and the mutual relationships between the different social and economic actors are well defined and, in time, effectively systemized. Industry delivers cheap and effective products and services capable of serving their clients' needs and wishes (while asking of them a minimum amount of physical and cognitive effort). In turn, the state does the same, offering the products and services that, for some reason, the private sector is unable to deliver (and following the same approach towards the minimum of requested efforts and participation). Finally, citizens work for a salary to be spent, in their free time, for buying the needed/wished products and services and/or to pay the taxes to access the services delivered by the state.

This idea of wellbeing, the system on which it is based and the behaviours it promotes can be discussed and deliberated from several standpoints. But what cannot be denied is the evidence of their present unsustainability. In fact, whether we like this idea or not, it is bringing us towards an environmental (and therefore also social) catastrophe: the dream of living better by consuming more clashes against the quantitative evidence that, when this same dream is dreamt by seven billion people today, and 10–11 billion in a few decades, it turns out to be a nightmare; on this planet there is simply not enough space, energy and materials for so many people desiring so much in terms of individual access to products and services (Vezzoli and Manzini, 2008).

It is on the basis of these observations that many authors have proposed a different vision of wellbeing: a sustainable wellbeing that, in principle, could be extended to the whole planet's population without major catastrophes. A wellbeing that is disconnected from the growth in the consumption and use of products and services. And, therefore, a wellbeing based on an idea that could be expressed in this way: living better by consuming less and regenerating the physical and social environment where we live. Finally, a sustainable wellbeing imagined not as a product to be consumed but as a condition to be collaboratively produced. And where new, sustainable qualities emerge, as the deepness of human relationships, of meaning of work, the pace of time and the quality of places become increasingly apparent and significant (Manzini and Tassinari, 2012).

EMPOWERING PEOPLE

The idea of a wellbeing disconnected from the notions and practices of owning, using and consuming products is not new at all. It was in 600 BC that Lao Tzu wrote the Tao TeChing, the basic text of Taoism, giving rise to one of its most famous quotations: 'Give a man a fish and you feed him for a day. Teach a man to fish and you feed him for a lifetime.' So the question of how to foster a non-product-based wellbeing had already been asked 2500 years ago and had found an answer that is as pertinent today as ever: to give people long-lasting wellbeing we must make it possible for them to deal with their own problems themselves. However, to do so they may need access to appropriate knowledge and tools; knowledge and tools that will enable them to do whatever they wish to do. As we have seen, Ivan Illich called them 'tools for conviviality'. We will refer to them here with the more technical expression of 'enabling systems'. To introduce them in the best way, we have to make a jump of 2600 years and move from China to Europe.

It was in 1998 when Amartya Sen, the great economist of Indian origin who had also been Dean of Trinity College, Cambridge, was awarded the

Nobel Memorial Prize in Economic Sciences. What is interesting for us here is that this prize was awarded to an author who deals with wellbeing and equity in such a way as to give some potentially very solid, useful theoretical foundations to our current discussion. What does Sen tell us? He says that what determines wellbeing is neither goods nor their characteristics, it is the possibility of doing things with those goods and characteristics (Nussbaum and Sen, 1993). It is this possibility that enables people to approach their own ideas of wellbeing and feel able to 'be' what they wish to be, and to 'do' what they wish to do (de Leonardis, 1994). In other words, Sen proposes a change in the way we treat well-being (and consequently the fair distribution of resources): instead of looking at people as bringing needs to satisfy (and therefore treating wellbeing as being a basket of products that addresses those needs), Sen proposes looking at human beings as actors who can, if they so wish and if they are enabled to do so, develop their capacity to act (which he calls capabilities) in order to achieve results (which he calls functionings).

I don't want to expound on Sen's theories here, save to point out how his approach could (and, in my opinion, should) form the basis of a new idea of wellbeing and of a design theory that could make it happen. Expanding people's capabilities is a very broad and rich perspective that can also be translated in very practical ways – especially when the discussion focuses on the contribution that social innovation can bring in respect of definition and diffusion.

In fact, what makes the discussion different from the ones we might have had centuries ago, or even decades ago, when Illich and Sen wrote their books, lies in the notion that social innovation, and specifically grassroots, community-based innovation is starting to emerge in the practice of a growing group of people, and not only in the theories of philosophers and economists.

FORMS OF COLLABORATION

Creative communities are based on action and collaboration. That is, they are groups of individuals who decide to link to each other in order 'to make something happen'. Therefore, we can say that the participants freely decide to reduce their individuality to create a system of links with other interested individuals. The way this collaboration takes place can be very diverse and very diversely motivated. In all the initiatives we are referring to in this chapter there is a blend of (1) discovery of the practical effectiveness of doing things together and (2) the cultural value of sharing ideas and projects. In contrast to what used to happen in traditional communities, this form of collaboration is not mandatory: it is collaboration by choice, where people can freely enter and, equally freely, make their egress.

The emerging active and collaborative behaviours are, however, not simply a step backwards towards old kinds of communities and traditional forms of collaboration. For example, the ways of living in the contemporary individualized and fluid society (Bauman, 1998) challenge the traditional communities and their forms of collaboration. Furthermore, new technological networks make room for new opportunities and, in particular, the use of the Internet and mobile technologies make collaborations easier and give social organization more flexibility. Above all, these emerging behaviours and collaborations provide innovative and original means to balance strong and weak social links (Granovetter, 1983). They give rise to a range of approaches for blending individualism and senses of commonalities vis-à-vis personal interests and social benefits (Sennet, 2012). In many ways, we could refer to the new forms of collaboration as technology-supported collaboration that permits contemporary, 'individualized' people to discover the power – and the pleasure – of thinking and acting together by combining collaboration and individuality.

To summarize, the emerging forms of collaboration enable people to find, case by case, the best compromise between individual freedom and collective endeavour. And in this way they result in potentially acceptable outcomes by both the declining Western middle class and the growing Eastern ones. They can also offer an attractive model for 'leapfrog strategies' in societies where traditional communities still exist. That is, in these societies they can indicate the concrete possibility to leap directly from dense and solid, and therefore rigid community-based collaboration to a flexible, technology-supported one.

Enabling strategies

Given their spontaneous nature, collaborative behaviours cannot be planned. Nevertheless, something can be done to make them more probable. When we consider ways to extend collaborative behaviours in different domains of everyday life, we see that this calls for a huge commitment in terms of time and personal dedication. Although this is one of the most fascinating characteristics of the original creative communities, it also potentially serves to limit their long-term existence and to increase the scale of the challenges associated with broader replication and adoption. So this appears to be the major limit to diffusion – that is, the limited number of people capable and willing to cross the threshold of commitment required to become one of the creative communities' promoters, or even just one of their active participants.

To overcome these problems collaborative behaviours need to become:

- more accessible (reducing the aforementioned threshold)
- more effective (increasing the ratio between results and required individual and social efforts)
- more attractive (enhancing people's motivation to be active).

Of course, for different collaborative behaviours to spread effectively specific enabling strategies are required. But it is possible, nevertheless, to identify some very general design guidelines. For instance, it will be necessary to: promote communication strategies that provide the required knowledge; support individual capabilities in order to make the organization accessible to a larger group of people; develop service and business models that match the economic and/or cultural interests of potential participants; reduce the amount of time and space required, and increase flexibility; facilitate community building; and so on. In turn, these strategies generate an articulated product-service system that, as a whole, can be defined as an 'enabling strategy'; a strategy that provides cognitive, technical and organizational instruments to enable individuals and/or communities to achieve a result, using their skills and abilities to the best advantage and, at the same time, regenerating the quality of living contexts in which they happen to exist (Jegou and Manzini, 2008).

What enabling strategies should do is to bring into play a specific intelligence – the intelligence that is needed to stimulate, develop and regenerate the ability and competence of those who promote and use them. Obviously, the more expert and motivated the user, the simpler the necessary solution may be. And vice versa: the less expert they are, the more the system must be able to make up for his or her lack of skill by supplying what he or she doesn't know or can't do. Finally, the less the user is motivated by economic and functional reasons, the more the system must be not only friendly, but also culturally attractive.

Spaces for Experiments

What can be done to implement these enabling strategies? When a strong grassroots initiative appears and seeks support from public agencies and/or other social and economic actors, the needed enabling strategy can be easily defined by listening to the explicit demands of the initiative and by interpreting its implicit needs (a well-known example is that of Community Gardens in New York, where, at a given point of their evolution, they started a positive collaboration with the City Council. This collaboration resulted in a new entity, the Green Thumb Agency, specifically designed to support them. Several other good, but less known examples could be considered and they would bring us to similar conclusions).

The situation is very different when there are no strong pre-existing initiatives. That is, when action should be taken by public agencies and other social and economic actors to trigger new local activities, to reinforce weak pre-existent initiatives or to spread good practices reproducing promising social inventions in new contexts. On the basis of the experiences carried out so far, it is clear that to make improvements to any of these strategies public agencies and other interested actors must create 'free spaces' where new solutions can be experimented (Manzini and Staszowski, 2012).

In fact, the systemic transformation in the relationships of different actors (and, sometimes, in their same nature) cannot be carried out in the ordinary conditions of their daily lives. In order to make such transformations possible it seems that some special circumstances have to be created: experimental spaces where different actors, civil servants included, can meet, interact, discuss different possibilities and enhance prototypes to verify them. This hypothesis, which outlines a very basic enabling strategy, emerges from personal experiences, but also – and mainly – is borne out of phenomenological observation.

Looking at what is happening globally, where active organizations are trying to promote radical innovations in the public realm, we can easily see that they have built, or are discussing how to build, these kinds of spaces. The spaces have different names (such as Living Labs, Fab Lab, Mind Lab, Change Lab, Solution Lab), but they also have a strong common denominator. As anticipated, they are free spaces where social and institutional experimentations can take place in an easier and safer way.

The experimentations that these spaces facilitate open two symmetrical opportunities: the opportunity for creative communities and hybrid social networks to move faster in their trajectory from the first heroic stage (the one of social inventions and their first prototypes) to the following stages of development where mature enterprises are created and, if necessary, dedicated, enabling products and services to be conceived and enhanced. At the same time, these experiment spaces offer to public agencies and other involved economic and social actors the possibility to enter into contact with citizens and their organizations in a context where, for them, it is easier and safer to test out new policies and new governance tools.

In short, in these experiment spaces it is possible to trigger and support a positive loop between bottom-up and top-down innovations. And, therefore, to promote the complex systemic innovations that are required to sustain and spread new ideas of sustainable wellbeing and the collaborative behaviours on which they are based.

CONCLUSIONS

The present economic crisis, combined with severe environmental and social challenges, is asking millions of people worldwide to search for new ways of being and doing. What will happen, in which direction they will search and where they will arrive at are open questions. Nevertheless, we can observe that several cases of social innovations are indicating some promising and viable directions. These directions, of course, are very diverse. But they have two common denominators: they are based on people's desire and capacity to be collaborative. They use the Internet and mobile technologies to make this collaboration easier and more effective. Being like that, their existence challenges one of the main trends of the past century – of becoming more and more passive and individualized consumers of goods and services. In fact, creative communities and hybrid social networks indicate the direction of active wellbeing: a sustainable wellbeing that is emerging worldwide but that, to spread, asks for and requires local and global systemic changes. And, therefore, for new design processes and new spaces for social and institutional experimentations.

NOTES

1. EMUDE is Emerging user demands for sustainable solutions: social innovation as a driver for technological and system innovation 2004–2006. The research consortium was coordinated by IDACO-Politecnico di Milano and included nine other international organizations.
2. CCSL has been an international research initiative, promoted by the United Nations. It started in 2008 and has been coordinated by INDACO-Politecnico di Milano and SDS-Brussels, where the EMUDE research results have been compared with similar cases in other contexts, in particular in China, India, Brazil and Africa.
3. DESIS is Design for Social Innovation for Sustainability. It is a network of design labs based in design schools (or in other design-oriented universities) promoting social innovation towards sustainability. These DESIS labs are teams of professors, researchers and students who orient their didactic and research activities towards starting and/or facilitating social innovation processes. Each lab develops projects and research on the basis of its own resources and possibilities and, at the same time, acts as the node of a wider network of similar labs, the DESIS Network, which enables them to exchange experiences and collaboratively develop larger design and research programmes.

REFERENCES

Baek, J.S. (2010), 'A socio-technical framework for collaborative services', PhD thesis, Politecnico di Milano, Milan.
Bauman, Z. (1998), *Globalization*, London: Sage Publications.

Bauwens, M. (2006), 'The political economy of peer production', *Post-Autistic Economics Review*, **37** (3), 33–44.
Bauwens, M. (2007), 'Peer to peer and human evolution', available at p2pfoundation.net, (accessed 8 January 2013).
Boyle, D. and R. Harris (2009), 'The challenge of co-production', available at http://neweconomics.org/publications/challenge-co-production (accessed 15 November 2012).
Boyle, D., J. Slaym and L. Stephens (2010), 'Public services inside out: putting co-production into practice', available at http://www.nesta.org.uk/library/documents/public-services-inside-out.pdf (accessed 15 November 2012).
Castells, M. (2012), *Networks of Outrage and Hope. Social Movements in the Inernet Age*, Cambridge: Polity Press.
De Leonardis, O. (1994), 'Approccio alle capacità fondamentali', in L. Balbo (ed.), *Friendly 94*, Milan: Anabasi, pp. 58–72.
Desis (2012), *Design for Social Innovation for Sustainability*, Presentation, available at http://www.desis-network.org (accessed 28 November 2012).
Granovetter, M. (1983), 'The strength of weak ties: a network theory revisited', *Sociological Theory*, **1**, 201–33.
Illich, I. (1973), *Tools for Conviviality*, Glasgow: Harper and Row.
Jegou, F. and E. Manzini (2008), *Collaborative Services: Social Innovation and Design for Sustainability*, Milan: Polidesign, available at http://www.sustainable-everyday.net/main/?page_id=26 (accessed 12 October 2012).
Joffé, G. (ed.) (2013), *North Africa's Arab Spring*, London: Routledge, Taylor and Francis Group.
Manzini, E. (2010), 'Small, Local, Open and Connected: design research topics in the age of networks and sustainability', *Journal of Design Strategies*, **4** (1), available at http://www.desis-network.org/papers/small-local-open-and-connected (accessed 8 January 2013).
Manzini, E. and E. Staszowski (2012), 'Public & collaborative', Internal document, Parson DESIS Lab, Design for Social Innovation for Sustainability, New York.
Manzini, E. and V. Tassinari (2012), 'Are sustainable social changes generating as new aesthetics?', internal document, DESIS Philosophy Papers, Design for Social Innovation for Sustainability, New York.
Meroni, A. (2007), 'Creative communities: people inventing sustainable ways of living', available at http://www.sustainable-everyday.net/main/?page_id=19 (accessed 4 January 2013).
Mulgan, J. (2006), 'Social innovation: what it is, why it matters, how it can be accelerated', available at http://www.youngfoundation.org/files/images/03_07_What_it_is__SAID_pdf (accessed 8 January 2013).
Nussbaum, M. and A. Sen (1993), *The Quality of Life*, Oxford: Oxford University Press.
Sennet, R. (2012), *Together*, New Haven, CT: Yale University Press.
Vezzoli C. and E. Manzini (2008), *Design for Environmental Sustainability*, Patronized United Nations Decade Education for Sustainable Development, London: Springer.

8. From energy policies to energy-related practices in France: the figure of the 'consumer citizen' as a normative compromise

Mathieu Brugidou and Isabelle Garabuau-Moussaoui

INTRODUCTION

The issue of public policies relating to changes in sustainable consumption has recently been raised in France and was a point of focus during the '*Grenelle de l'Environnement*', a consultancy and decision-making system bringing together, for the first time, environmental non-governmental organizations (NGOs), companies, trade unions, members of parliament and government officials.[1] However, since the 1970s, France has been faced with the issue of behavioural changes in energy consumption. This chapter examines the question of the behavioural changes driven by energy-saving policies in France.

In particular, the recent history of energy-saving policies has led to the emergence of the 'consumer citizen' in relation to market-framed policy devices. Yet analysis of certain public policies (using energy-saving light bulbs for lighting, thermal renovation via the sustainable development tax credit) shows that while frameworks for consumption behaviour can be quite effective when it comes to designing policy, there are shifts and 'spill-over effects' in the interlinking and confrontation between the figure of the 'consumer citizen' and the actual practices of households and of all the actors involved. Finally, energy practices in French households demonstrate a multitude of social logics either favouring or restricting energy-saving practices, both within 'concerned' social groups and amongst the 'general public', where the environmental argument is far from being a pertinent indicator of 'energy care'. A better understanding of the social mechanisms at work in energy practices and norms would provide new fulcra for public policies, which are too often based on a conception of homo economicus/

ecologicus, designed to maximize his financial and environmental utility, with no consideration for the contradictory imperative embedded therein.

HISTORY OF FRENCH ENERGY-SAVING POLICIES: THE EVOLUTION OF THE END USER

Introduced during the first oil crisis in the 1970s, energy-saving policies are nothing new in France. They have been subject to the ebb and flow of numerous development cycles. There have been four main periods, each highlighting different types of end user: depending on the period, end users would appear to be more or less central to energy policy tools and they are endowed with different characteristics (service user, product user, consumer and citizen). In this section we show how public energy policies have approached the issue of end users and their changing behaviour.

First Oil Crisis 1974–79: Two Public Policies?

The first oil crisis led to two types of public action, one directed towards demand (reducing the amount of energy used by businesses and households) and the other towards supply (the launch of an ambitious nuclear-based electricity production programme). These two policies were to some extent linked, with energy savings providing an interim solution until the nuclear power plant construction programme reached maturity. But they might also appear to offer contradictory visions (thrift versus production) for opponents of nuclear power. The first oil crisis was an opportunity to implement various informative, incentive and regulatory measures. The Agence pour les Economies d'Energie[2] (AEE), created to implement the energy-saving policy, thus launched the first campaigns targeting the general public and portrayed the figure of the 'responsible service user'.[3] As the Agency's director, Jean Syrota declared the purpose of these campaigns was to tone down the notion of constraint that was so present: 'On the contrary, we wanted to reassure. . . . The basic idea was that we had to treat the French people as being intelligent, reliable and responsible' (quoted by Pautard, 2009, pp. 34–5). The campaigns also introduced the first 'energy labels' designed to inform consumers about the energy consumption of certain electrical appliances.[4] Finally, regulations defined modalities for sharing electricity in case of shortage and the first thermal norms for collective housing. The creation of summer and winter hours was an opportunity to combine regulatory measures with information campaigns.[5] At that time, the rhetoric on responsibility was already camouflaging upstream control over behaviour.

Second Oil Crisis (1979–83): Contradictory or Interlinked Policies?

The second oil crisis (1979–83) was an opportunity to accentuate certain aspects of these policies. The call for responsible citizenship was highly dramatized: 'If consumers do not demonstrate an exceptional spirit of responsibility, the world could well be heading for disaster'[6] declared French Prime Minister Raymond Barre. Yet these dramatics also related to the nuclear power programme, which offered the opportunity to script the citizenship aspect of end users. With regard to the development of the French nuclear power programme, the then French President Valery Giscard d'Estaing said: 'It's a little bit as if the French people had with their own hands built an oil field capable of producing, on an annual basis, more than half of what our British neighbours pump out of the North Sea, but an oil field which will never run dry'.[7] The level of investment aid for thermal improvement projects was increased fourfold, and the AEE's resources were substantially increased, allowing the agency to gain a foothold throughout France. The political changeover that brought the socialists into power in 1981 had no effect on these policies: the nuclear plant construction programme continued, despite F. Mitterand's initial undertakings (Blanchard, 2009). National electricity supplier EDF developed a programme promoting electric heating. This programme can be seen as a way of combining the demand-side management policy with the nuclear production policy, with a view to replacing fossil fuels, but it came under criticism, with people feeling that a policy for an abundant and relatively cheap offering contradicted consumption reduction objectives (Hecht, 2004).

In 1982, the AFME (French agency for energy management) replaced the AEE,[8] leading to two significant changes. This was the chance to focus on an energy 'control' programme oriented towards technical research on energy efficiency and performance, such a technico-economic perspective bringing product users into the limelight (the ergonomics of systems, the average perceived comfort temperature and so on). Furthermore, it increased decentralization by delegating part of the job of informing and raising awareness to local levels. French Prime Minister Pierre Mauroy stressed the need for dialogue and consultation at local level, integrating associative dynamics. It was a case of 'allowing new citizenship to also be exerted in the field of energy, with information being provided to one and all and a true sharing of responsibilities between local authorities, regions and the State'.[9] In fact, the socialist party was divided on this issue, between production-oriented and anti-production sentiments. Governmental arbitrations and discourses reflected these tensions.

The Counter Oil Crisis (1984–96): Energy Savings Overshadowed

In the middle of the 1980s, the sharp drop in oil prices and the increasing number of nuclear power plants connected to the network[10] were to lead to a major change in public energy policies, in particular within a political context swinging to the right. This in turn led to a dwindling of the figure of the citizen, in favour of the threefold figure of 'efficient equipment', the product user and the consumer (in the sense of managing a budget and choosing goods in a market). The AFME, which was allocated reduced resources, significantly changed its strategy: it was henceforth a question of backing research and development (R&D) for effective equipment with which to build an efficient market. Agreements had to be made with household appliance manufacturers in order to help consumers who were somewhat in the dark to calculate their energy costs. In 1992, the AFME became the ADEME[11] (French agency for the environment and energy management), which in addition to dealing with air pollution and waste, was tasked with concentrating on renewable energies and co-generation. 'Behaviours' were interpreted through the historical energy provider's pricing policies (demand adjusted to suit offering to a large extent structured by the characteristics of the nuclear power stations). The different plans for boosting the economy through household consumption, within the context of an abundant offering, left little room for energy-saving policies. These changes went hand in hand with a significant increase in end user energy consumption.[12]

The Current Emergence of Environmental Issues, the Internationalization of Public Action and Deregulation

The final period saw the emergence of new environmental issues (climate change in particular), the internationalization of public energy policies (due in particular to Europe) and the deregulation of the energy sector. France's commitment to not increase the level of its carbon dioxide (CO_2) emissions under the Kyoto protocol in fact contributed towards a redefinition of energy management policies. A new thermal regulation (RT2000) was voted. A communication plan linking energy and the environment was implemented in parallel to the creation of a wider *Points Info Energie*[13] network aimed at the general public. Energy policies began to be decentralized.[14] In 2004, the Energy Orientation Law[15] (LOE) was passed, setting out new energy policy priorities: to combine continued energy security, protection of the environment and the maintenance of 'competitive' prices that, in particular, would give all citizens access to energy. Energy management, the development of renewable energies, the construction of a new-generation pressurized water nuclear reactor (EPR) and backing for research were all among the government's priorities.

The deregulation of the sector as from the 2000s,[16] driven by European policies, changed the modalities of public action. The diversity of the targeted publics and objectives led to an increase in the number of instruments and to the creation of market instruments covering matters such as CO_2 emission quotas or energy-saving certificates (Pautard, 2009). With regard to the energy market opening up to the general public (July 2007), the CRE[17] (French energy regulation commission) was tasked with ensuring that the French electricity and gas market functioned correctly and with establishing the figure of the consumer in this sector. This insistence on the self-regulation of the marketplace went hand in hand with a reduction in ADEME's allocated budget.

Finally, as from 2007, the '*Grenelle de l'Environnement*' was determined to develop a decisive turning point in public policies relating to the environment. Although the system received mitigated approval – particularly if we consider the lengthy processes of public policy (Boy et al., 2012), it demonstrated the return to the agenda of energy policies and of the issue of energy saving in a brand new form. Indeed, the *Grenelle* set out a strong link between environmental issues (especially climatic) and the market. It deployed information and incentive devices within both the public arena and the marketplace and paid considerable attention to the figure of the 'consumer citizen', on the one hand, debating energy choices from the standpoint of ecological democracy, and, on the other hand, opting for 'sustainable consumption' in line with economic and environmental considerations. Such consultation was, however, only possible at the price of preliminary negotiations that excluded the nuclear issue.

Energy Policies as Part of a Global Transformation of Public Action

This brief summary of public policies in the field of energy saving highlights a succession of configurations of public action that differed considerably due to outside constraints (oil crisis, awareness of the depletion of fossil fuels, climate change and so on), the mounting influence of international (European Union, Kyoto agreements) and local bodies and decisions, and political logics. However, some of the changes affecting governance of the energy sector were similar to those affecting all of public action in France: government intervention was increasingly framed between the local level – the result of decentralization – and the European level (Borraz and Guiraudon, 2008). This context led to the emergence of independent administrative authorities and to a more vigorous associative democracy (Christiansen and Kirchner, 2000).[18] For example, the Commission Nationale du Débat Public (CNDP),[19] created in 1995 and reformed first in 2002 and later in 2010, 'is part of a more global movement to strengthen information instruments and participation in the policy of public risk management, these instruments having a dual vocation, that of the democratisation of decision-making and that of the orientation of citizen behaviour' (Lascoumes, 2008, pp. 49–50). The increasing recourse to these instruments was also a response to the objective of informing populations, making them aware and responsible, in order to encourage them to change their behaviour: 'Policies over the last decade relating to the recycling of household waste and those now targeting saving water and energy, are to a very large extent based on accountability through information' (Lascoumes, 2008, p. 50).

At the same time, these public policies were increasingly characterized by market instruments at the service of greater individualization, of 'focusing on individual behaviour' (Borraz and Guiraudon, 2010, p. 15). We saw that these combinations of instruments defined different but partially connected figures of end users: during the first oil crisis, the figure under the spotlight combined the citizen with the energy service user. In the 1980s, the product user then came into the picture, in a context of the public monopoly of energy supply, of the deployment of 'efficient' technical devices and of a decrease in concerns about energy. Progressively equipped with different types of commodity, the consumer gradually broke free, liberated from these civic concerns (that is, energy savings) by the virtues of energy efficiency – that transformed the issue of frugality into one of cost. Finally, a 'consumer citizen' appeared, linking environmental and market stakes within a framework of sustainable development. Furthermore, citizens continued to express themselves at local (in relation to infrastructures, for example) or national (for example, via the *Grenelle de l'Environnement*) levels.

EXAMPLES OF PUBLIC POLICY INSTRUMENTS RELATING TO ENERGY SAVINGS: THE MANAGEMENT AND APPROPRIATION OF DEVICES

The government and the market thus developed tools, instruments, products and infrastructures, which we might generically refer to as 'devices', to provide a framework for energy-consuming behaviour and to construct the figure of the 'consumer citizen'. This section examines two public policy devices in order to show how the behaviour-framing process works in practice and to demonstrate how these devices are 'translated' (Akrich et al., 2006) by the different actors (state, market, civil society, households) involved in their implementation. The devices are thus jointly constructed, reinterpreted or even 'spilled over', depending on each individual actor's strategies and constraints, and on their relations with one another. The two devices we present were chosen in accordance with various criteria: the diversity of the type of instrument (information plus market, then constraint, versus tax incentive plus market), the diversity of the type of behaviour (gesture versus investment) and the availability of relevant studies.

Lighting and Energy-saving Light Bulbs

Lighting is an item of consumption appropriated by both public policies and households in order to achieve 'energy savings'. By cross-referencing the history of public policies on lighting and their distribution among the general public with that of actors in the lighting market and that of social practices and their transmission, we can see where the frameworks and translations proposed and performed by the various actors are situated.

'Visible' consumption and savings

First and foremost, one might wonder why lighting has become a symbol for energy saving, when this item's share of a household's overall consumption is relatively small (less than 2 per cent of domestic energy costs (Besson, 2008).

Furthermore, the implementation of a public policy regarding lighting and appliances practices is particularly complex, as with each household having an average of 25 light bulbs, it involves millions of light bulb changes and little gestures. However, households consider that lighting is one of the only 'visible' sources of energy saving, having the impression that they can see the electricity being used in the lighting, just as they can see water coming out of taps (Moussaoui and Vaubourg, 2004). In addition, the energy-saving awareness campaigns of the 1970s, related to lighting, were 'embodied' by children and adults at that time and then passed on to later generations (Garabuau-Moussaoui et al., 2009). Lighting also has very 'strong' social functions (comfort, ambiance, safety, territory, riches and so on), which are on the face of it incompatible with the notion of savings. The contradictory imperative is 'resolved' in the action of 'switching off when you leave the room', something that avoids one having to question either the electricity consumption when members of the family are in the room or the power or number of light bulbs (Moussaoui, 2007).

Improved unitary energy efficiency, but slow diffusion and no slow-down in the acquisition of multiple appliances

Evolution in the way people light the living rooms is tending to shift from a central light source (on the ceiling or in the form of a halogen lamp) to multiple points of light (or even string lights). So 'in the space of 25 years, the French people have trebled their consumption by increasing the number of light sources (330 million light bulbs are sold every year).'[20]

At the same time, 'energy-saving' light bulbs are seeing very limited success. In 2006, the market share of compact fluorescent lights (CFL) was modest (6 per cent of sales): 'In 1998, 25 per cent of households said they owned one; 35 per cent in 1999 and 44 per cent in 2001' (Zélem, 2010, p. 39) with 68 per cent in 2008. Given that each household has an average of almost 25 light sources, having an energy-saving light bulb remains a highly 'symbolic' act and it wasn't until 2005 that a quantitative survey asked households if they equipped 'a large share of their lights with energy-saving bulbs'. Only 15 per cent gave a positive response (Roy, 2006). During the same period, there were an increasing number of operations to distribute low-energy light bulbs or to sell them at reduced prices.

Studies on the use of these light bulbs (Moussaoui and Vaubourg, 2004) show that they are used in places of little social consequence: toilets, corridors, cellars, garages and so on, energy-saving light bulbs are in fact considered to be expensive and to give off light that is not very pleasant ('cold colour', 'pale'), taking time to light up and being unaesthetic (their shape does not suit the light sources purchased). They are therefore 'stigmatizing'. These disadvantages become advantages in children's bedrooms: on bedside tables there is no longer any worry that children will burn themselves; they also serve as night lights for children who are afraid of the dark. In the middle of the 2000s, although the general public were more aware of them, energy-saving light bulbs were still not used to any great extent in people's homes, despite promotional campaigns and incentives to buy them based on two arguments – cost and ecology: 'A light bulb which reduces your bill and preserves the planet? Yes, it's possible'[21] (ADEME campaign, 2008).

From encouragement to regulatory constraint: building a process to bring market actors on board

It is within such a context that several European directives are attempting to reduce energy consumption for lighting (Catteau et al., 2010).[22] The laws initiated by the *Grenelle de l'Environnement* confirmed these different regulations and have made it possible to bring forward to the end of

2012 the deadlines planned in European calendars for replacing incandescent light bulbs with 'efficient' bulbs, by signing a voluntary agreement between public authorities and actors in the energy sector.

The obligation to replace incandescent light bulbs with energy-saving light bulb has conducted the state to structure the market, but also and above all to propose some incentive devices for the actors involved, so that they would play the game and plan ahead to meet European obligations.

Certain companies had taken the initiative regarding this regulatory evolution and had already turned constraint into opportunity. Very early on, light bulb manufacturer Philips had oriented its research and investment strategy towards so-called 'eco-responsible' solutions: it used lobbying to put these issues onto the European political agenda and participated in discussions at the *Grenelle de l'Environnement*. Its innovation strategy was a way of getting ahead of its Chinese competitors, who would not have the resources to develop the new technologies that Philips was offering. Furthermore, Philips and other manufacturers are currently orienting their research towards light-emitting diodes (LEDs), with CFLs being considered as a transition technology (Catteau et al., 2010). Here, new regulations can thus be seen to support industrial strategy.

Manufacturers and retailers found that these 'incentive' devices could be integrated into their commercial strategies and competitive positioning. They 'translated' the incentive devices into devices aimed at the general public (compatible with the figure of the citizen-consumer: the fight against climate change, financial savings, long life, child safety and so on) and modified their products and packaging to suit these advertising messages (Pautard, 2009). It is striking to note that 'civil society' has taken the same direction, with the majority of consumer and environmental associations helping the distribution of energy-saving light bulbs rather than hindering it, despite the controversies that have received little attention (mercury, electromagnetic fields).

Awareness campaigns had a role to play in achieving acceptance of the constraint. Unlike other countries,[23] the French people would appear to have 'accepted' this obligation to replace bulbs. Is this because they have not yet changed them all? Or is it because public policies and the marketplace have helped create a 'norm' that is in the process of being integrated? Campaigns have now switched focus onto the recycling of light bulbs, backed by the creation of the Recyclum organization, and the issue now is the shift from purchase to recycling (use being left unspoken). A new paradox has appeared: while the various actors have come to a consensus to hide the mercury controversy, recyclers would gain from using this argument to 'mobilize' consumer citizens.

Case study conclusion

This example shows that the construction of a choice for consumers, first between incandescent light bulbs and energy-saving light bulbs, then – nowadays – between several energy-efficient technologies, is managed by the authorities, in partnership with manufacturers, retailers and civil society. The obligation to replace has not therefore led to market actors having less influence or to the end of information campaigns and the enrolment of actors. Public authorities have looked both to provide incentives for market actors and to equip consumers (energy labels, promotional campaigns and incentives to buy, followed by collection systems for recycling). The state here plays the role of incentive provider (in particular, by bringing the calendar forward) and regulator, but also has the role of constraining (regulations). In this example, as in others (Calwell, 2010), the question is whether or not the strategy of energy efficiency per product (which would seem to be one of the avenues of consensus between state and the market), when it is not coupled with a strategy of energy moderation and frugality (no public policy on the use of appliances) can meet the environmental objectives upon which the directives and regulations are founded.

The Thermal Renovation of Buildings: The Sustainable Development Tax Credit Example

Designed to improve energy efficiency in buildings, the Sustainable Development Tax Credit[24] (SDTC) introduced in 2005 has been a great success. This tax incentive was designed to reach all households, with those who pay tax seeing their tax bill reduced, and those who do not pay tax receiving a cheque.[25] Over the first four years, it led to 5.4 million renovations for a total of €23.6 billion, and concerned one out of seven main residences. However, analyses by Vauglin et al. (2011) have shown that this success encountered certain limits: first, it would seem that the government found it hard to predict how much this measure would cost:[26] its annual cost increased significantly, from €1 billion in 2005 to €2.5 billion in 2008, which led the government to propose a far less attractive measure, that of the interest-free eco-loan.[27] Furthermore, only some categories of the targeted population seem to have been interested, carrying out types of work for which the energy-efficiency objective was secondary. Generally speaking, it was in fact well-off households, homeowners and residents in detached houses who benefited from a measure that they saw as a tax opportunity ('deadweight effect', said the experts interviewed by Vauglin). Vauglin et al. have highlighted appropriation strategies employed by different types of actors that partly explain these 'failures'. Households had the work done not so much for 'environmental' reasons – improving the building's energy efficiency – but rather for reasons of comfort, economic value or to renew their installations. They thus preferred to change windows or the boiler, when '[technico-economic] rationality would suggest starting with loft or wall insulation' (Vauglin et al., 2011, p. 18).

Yet this selective appropriation by households would not have had such an effect if it had not combined with the economic logic of certain business sectors: the industrial windows sector, for example, 80 per cent of whose market was concerned by the SDTC, actively strove to 'harness' this measure. Finally, the government would appear to have been using the measure to pursue several objectives, some of which were divergent: in addition to the environmental aim, it wanted to shore up the economy (by supporting business sectors[28]) and reduce costs. Furthermore, the government's difficulty not only in predicting distribution of the measure but also refocusing its design in accordance with the modalities of its appropriation by both households and business sectors – which supposed careful monitoring of the measure – undoubtedly explains some of the SDTC's limits.

Like all innovations, policy instruments directed at the marketplace configure forms of consumption that are themselves reformulated by professional actors and end users. We know that an innovation's success depends in part on its capacity to integrate these appropriation logics. Yet in this case, the state would appear to have had certain difficulties in 'renegotiating' the 'use script' (Akrich, 1992) embedded in the instrument when taking into consideration the reformulations of the different actors.

In conclusion, these two examples show that consumer behaviour is shaped by public policy instruments, but that there are spill-overs, strongly supported by market frameworks, linked to the diverse motivations of consumers who reappropriate these measures, 'translate' them and find a meaning for them within their practices (comfort, renewal of household equipment and also the environment) in accordance with numerous social logics.

FROM CHANGE IN BEHAVIOUR TO AN UNDERSTANDING OF THE DYNAMICS OF PRACTICES AND SOCIAL NORMS

If we look at this from the standpoint of households, the figure of the 'consumer citizen' that is currently being drawn in energy policies remains a fragile construction. Indeed, public discourses maintain an ambiguity by voluntarily deploring the lack of environmental commitment from 'consumer citizens'. Even when they have been 'made aware', households do not appear to take concrete action. But as we have seen, this observation is largely due to highly simplified (or even minimal) representations of use as they are encoded within the state's instruments: the civic failures that are so deplored very often camouflage the inability of these instruments to transform and reduce the diversity of uses. But, in our opinion, a household's commitment to energy efficiency or frugality devices depends more on a capacity to appropriate than on environmental values. In particular, the notion of energy care is currently being developed at the intersection of social practices and norms, of public policies, of debates and controversies, of commercial offerings and socio-technical devices. This does not necessarily mean that households are saving more energy than in the past, but that we are seeing the constitution of an energy-care practice (including a repertoire of shared actions, skills and meanings (Warde, 2005) and the emergence of a new social norm (bringing together a set of devices, practices and justifications for practices) in mainstream households.

This reconfiguration of practices and social norms is linked to an evolution in the discourse on the connection between energy and the environment, to political and commercial instruments and devices (see above) and also to a dynamic of possible consumption customs and politicization on the part of end users, which we analyse using Hirschman's trio (1970), helping to connect public and private concerns, political and market dimensions, and their social dynamics:

- The energy-care norm might be the object of multiple reappropriations; it will then be a case of paying attention to the description of the diversity of the practices structured by the norm ('loyalty').
- It might also be confronted with practices firmly structured by other norms – such as comfort – and fail to reconfigure them in accordance with the new normative regime ('exit').
- Finally, it might give rise to a politicization that, for example, condemns its modalities of application or its founding values ('voice').

Loyalty: The Normalization of Practices

The many works of research on households' energy practices[29] have shown that energy care has been caught up in and overtaken by a consumption system, which explains the high levels of consumption. An extremely weak link has therefore been established between environmental attitudes and pro-environment actions, in particular through analysis of the relationship between energy consumption and income (Dobré, 2002; Wallenborn and Dozzi, 2007). Similarly, the notion of 'compartmentalization' (Halkier,

2001) of areas of consumption (food, leisure, mobility, clothing and so on) helps us to understand that these areas of life do not involve the same socio-technical devices, life priorities, symbolic meanings and so on, leading to heterogeneous practices and norms. This phenomenon is found in items of energy consumption – lighting, heating, cooking and cleanliness refer to different practices, social representations, equipment and public policies.

However, despite this diversity, we were able to identify several recurrent energy-care logics (Garabuau-Moussaoui, 2011). An analysis by generation and by life stages allowed us to reveal historical 'trends' for the development or, on the contrary, for the minimization of these logics, very similar to those seen in this chapter:

- A logic of anti-waste among the most elderly (who lived through the war, the rationing or the ensuing shortages) passed down from one generation to the next, which still remains a benchmark, even for children today.
- A financial logic, which is very strong in every generation because it is the simplest way to translate energy conservation efforts (smaller bills are a visible sign of decreasing consumption) and because budget management ('domestic economy') is a vital part of domestic activity for all generations.
- A rising environmental logic, which is hard to accept if it clashes with financial logic (for example, reluctance to pay for energy-efficiency services, for energy management/conservation advice or for organic products).
- A logic of criticizing overconsumption, which is re-emerging today (it first appeared in the 1970s).
- A regulatory logic, some people being sensitive to the fact that certain practices are 'required' by technical devices, laws, the government's awareness campaigns (sorting waste, turning off appliances' standby mode and so on).

These configurations of practices and discourse, given structure by the energy-care norm, can be accompanied by normative 'police' work and by a stigmatization of deviant behaviour (Becker, 1963), in relation to the figure who 'thinks that it is not important to save energy', backed by terms such as laziness, stupidity, selfishness or greediness, on the basis that such people do not 'need' or do not 'want' to be careful because they are only interested in 'personal' and individualistic comfort, whereas those who are careful show 'concern' towards the planet and towards other people (Brugidou, 2011).

Exit: Normative Conflicts

These energy-care logics can be counterbalanced by other logics. In particular, the comfort logic is a very strong social norm that for the most part remains linked with material comfort. This logic has generally gone unquestioned and has increased since the Second World War. For the people interviewed, there are almost never any doubts about accumulating energy-consuming electronic devices and appliances. They consider it to be 'normal' that the 'modern' standard of living (Shove, 2003) includes a 'material culture' of energy and appliances that consume a high amount of energy.

Furthermore, many logics are ambivalent, that is, they can lean towards energy care or towards energy consumption. In this sense, parents can be confronted with contradictory imperatives, that of educating their children to be careful with energy and that of providing them with comfortable living conditions (sometimes by sacrificing their own comfort). Should one turn up the heating in the children's bedroom or should one keep the temperature down for health reasons? Should they be given a bath every day? Should one buy the game console they are asking for?

Similarly, social distinction, for so long based on the accumulation of material goods, is in certain milieus being transformed into 'ostentatious sobriety' (Moussaoui, 2007). Hybrid cars, solar panels, heat pumps and wood-burning fires are becoming meaningful identity markers.

These different logics of action come together and evolve over time. The integration of a new element in the everyday world might therefore create chain reactions relating to the dominant logics of action. This is referred to as the Diderot effect (Shove and Warde, 1998) and the rebound effect (Gossart, 2010) when it leads to a rise in consumption, but there can also be 'trajectories' oriented towards greater energy moderation. However, compartmentalization can prevent or minimize this process of distributing new practices.

Voice: Politicization of the Norm

Voice – in the sense of politicizing an issue – is the typical mode of action used by the 'moral crusader' (Becker, 1963), who wishes to introduce a new norm based on values and on a representation of the common good. This is why the instigation of debates on environment and energy-related issues in numerous arenas (media, social and personal networks and so on) contributes towards the evolution of collective norms, diffusing the arguments and controversies into mainstream discourses. It would also increasingly appear to be a required route for the construction of public policies. With regard to energy saving in France, in its 2007 'Quick, it's getting warm'[30] campaign, the ADEME included a public debate device on demand-side management (Benvegnu, 2011) and the *Grenelle de l'Environnement* was presented by its promoters as a process of consultation between the different stakeholders that included a phase of consultation with the general public (Boy et al., 2012). Everything took place as if these devices were playing the role of legitimizing (or even ritualizing) (Lassègue, 2010) public policies for the purpose of verifying the 'acceptability' of the proposed measures.

Consumer citizens, while enjoined to give their opinion, relied on different forms of authority (experts, political representatives, interested groups or publics and so on) who guaranteed these proposals, thus exonerating them from having to go through the cognitive and affective processes that would allow them to construct a belief. So for the majority of citizens, the appropriation and acceptance of a public policy do not require an ideological commitment or a belief (the prerogative of 'moral crusaders', mobilized groups or concerned publics – Brugidou, 2008), but rather a relative detachment, in a balance between 'concern' and confidence/delegation in the devices implemented to deal with the issues.

However, the process of institutionalizing a norm and its distribution is not a linear one. The phasing-in of a norm, through implementation of public policy devices, very often gives rise to periods of controversy[31] that can undermine the institution of the norm ('what is women's role in new ecological practices?' asks Badinter (2010), and 'what about a degrowth model, rather than a green growth model?' say Degrowth militants (Latouche, 2006) and the Economistes Atterrés (2012). These controversies can also relate to the ways in which a norm is applied. For example, the waste-sorting norm went through a period of strong development that began when a collection (separate bins and collection by the authorities) and recovery system was introduced, demonstrating 'the need for an institution which imposes its rules in order for individuals to self-impose new obligations' (Dujin et al., 2007, p. 85). But the analysis of the discourses of French people who answered a quantitative survey on sorting waste (Brugidou, 2011) shows that while the principle of sorting is not called into question, the ways in which the norm is put into practice – through criticism of how the collection is organized, that is, implementation of the socio-technical and regulatory devices – can be the subject of controversy: responsibility for the absence of any sorting is placed on the local authority's failure to implement the norm. In the same way, the carbon tax gave rise to major controversies in 2010 during the *Grenelle de l'Environnement*.[32] Such controversies can go hand in hand with a norm's distribution (to whatever degree) within the social space; from this standpoint, the waste-sorting norm would seem to be more widely implemented in France than is energy saving (Brugidou and Moine, 2010).

The different studies carried out by ourselves and by other researchers show that energy-care practices do exist, that households are 'careful' about energy (albeit with varying conceptions) and that they are prepared to 'accept' new devices for energy efficiency, energy moderation and renewable energy production, if they can appropriate them and if they can integrate them into the system of consumption that they construct from one day to the next. This takes place when household practices are faced with a 'transformation challenge', through the appropriation of information or a device within a broader system of consumption and life project.

So households are not torn solely between a 'consumer' identity and a 'citizen' identity. They are dealing with a whole set of issues and cannot simply 'vote with their wallets'. This does not remove the firmly rooted existence of energy-care practices, which take place in the name of various different (but not infinite) logics, demonstrating the social reinterpretations of public policies.

Integrated practices and the institutionalization of norms demonstrate both the constitution of 'energy care' and the maintaining of 'comfort', even when criticized. Yet far from opposing, on the one hand, public policies that are virtuous in their attempts to achieve the advent of an energy transition, with, on the other hand, households who are bad citizens (Rochefort, 2007) resisting change, this pairing is embedded within the current 'socio-technical regime' of energy (Geels, 2002), both in France and in other European countries (Fudge, 2010). Indeed, we find important occurrences of these two themes in the different spheres of influence and of construction of social practices relating to energy. As far as the political sphere is concerned, we see the emergence and development of public environmental/energy policies and in particular the deployment of end user awareness and market instruments (the symbolic and economic management of the current political situation) and, at the same time, the development of a discourse on the need for growth in times of crisis (which requires cheap and abundant energy), and thus for consumption (political management of the current economic situation). Similarly, in the market sphere, there is both the strong influence of public policies (market instruments) – along with pressure from consumers to 'greenify' and for companies to communicate – and strong consumerist pressure (Rumpala, 2009).

CONCLUSION

The 'consumer citizen' is therefore a 'policy' figure (in the sense that he or she is constructed during the implementation of a public policy) who tries to reconcile comfort and the environment in relation to both energy issues and, more generally, consumption issues. The state and the marketplace rigidly structure this figure, both by developing a discourse of responsibility and of the consumer-actor, intended to 'individualize' the agent, even in his or her sense of citizenship (making a small gesture to help the planet, voting with his or her wallet), and by implementing devices that effectively orientate actions in the desired direction (smart grids, nudges and so on).

This figure constitutes a normative compromise for several reasons. First, the market and regulatory devices that implement it relate to different norms: to varying extents they favour comfort or the environment, a pairing that is forever under tension. Furthermore, they are conveyed by networks of actors that promote different solutions (technology versus behaviour, efficiency versus sufficiency). Finally, they are the result of an accumulation, or even of sedimentation, of the historical strata of public policies (Hamelin and Spenlehauer, 2008). The processes of institutionalizing norms are more or less continuous, with periods of intense production, stagnation or even relative reversals. For example, one might say that France's energy-saving norm of the 1970s and 1980s was partially 'uninstalled' (to use a software term), victim of a process of deinstitutionalization, with the discursive and regulatory devices of energy-saving policies falling somewhat by the wayside (thermal regulations for buildings not being implemented (or not fully), awareness campaigns being further and further apart and so on).

We must distinguish between the definition of the 'problem' of energy saving, which would currently appear to favour the 'causal story' (Stone, 1989) of the 'citizen consumer', and the range of solutions that are actually implemented, combining extremely varied instruments (Hamelin and Spenlehauer, 2008), because they result from these processes of sedimentation and compromise and from the confrontation between policies and uses (rejection, appropriation, politicization).

NOTES

1. For a presentation in English, see http://www.legrenelle-environnement.fr/-Version-anglaise-.html?rubrique33 (accessed 7 July 2012).
2. French energy-saving agency.
3. From 1946 to 2004, EDF was a public service utility and had the monopoly of the production and commercialization of electricity. So households were 'public service users'. Progressively, new optional tariffs were initiated by EDF, transforming users into consumers. Since 2004, market liberalization and EDF's statutory change from public to private status have achieved this shift.
4. Following a voluntary agreement with manufacturers, 20 years before the European initiative (Pautard, 2009).

5. Journalists were shown how the one-hour time shift would be implemented throughout the country.
6. Extract from a speech by Prime Minister Raymond Barre following the second oil crisis; quoted by Leray and de La Roncière (2003, p. 26).
7. Quoted by Pautard (2009, p. 47).
8. In 1982, the Agence Française pour la Maîtrise de l'Energie (AFME) had a budget of FF 700 million and 250 employees. In 1983, during F. Mitterrand's first presidency, its budget stood at FF 1.3 billion. It employed 500 staff spread throughout France.
9. Quoted by Pautard (2009, p. 53).
10. The nuclear production capacities set up in 1986 became greater than the capacities currently being built or in project form.
11. Agence de l'Environnement et de la Maîtrise de l'énergie.
12. Total final energy consumption only increased by 0.4 per cent between 1973 and 1980, but saw a sharp 6 per cent rise in the 1980s, followed by a 12 per cent increase in the 1990s (Meuric, 2006). Improved energy efficiency per appliance did not therefore curb the propensity to buy more appliances.
13. Energy information points.
14. Decree relating to the *Schéma de Services Collectifs de l'Energie* (community energy services scheme) enacted in April 2002.
15. Loi d'Orientation sur l'Energie.
16. French law no. 2000–108 dated 10 February 2000 relating to the modernization and development of the public electricity service, French law no. 2003–8 dated 3 January 2003 relating to gas and electricity markets and to the public energy service, French law no. 2010–1488 dated 7 December 2010 on the reorganization of the electricity market (NOME).
17. Commission de Régulation de l'Energie.
18. For a discussion on this concept, see Saurugger (2003); Grossman and Saurugger (2006); and Jobert (1999).
19. French national commission for public debate. http://www.debatpublic.fr, accessed 7 July, 2012.
20. http://www.arehn.asso.fr/dossiers/ampoules/ampoules.html (accessed 7 July 2012).
21. In French, *Choisir une ampoule qui réduit sa facture et préserve sa planète, aujourd'hui, on peut.*
22. Directive no. 2005/32/CE, complemented by regulation no. 244/2009, aims to 'integrate energy saving when appliances are designed'. In addition, an earlier directive was introduced to oblige 'manufacturers and importers to fund the collection and processing of used appliances' (Directive 2002/96/CE, brought into French law by decree no. 2005–829 dated 20 July 2005). A calendar for replacement is set out in European regulation 244/2009.
23. See the controversies in the USA, triggered by the Tea Party and taken up by the Republican Party (http://www.goodplanet.info/Contenu/News/USA-les-republicains-declarent-la-guerre-aux-ampoules-a-basse-consommation) or the incandescent bulb stockpiling phenomena in Great Britain, Germany, Austria, Hungary and Poland (Cateau et al., 2010).
24. *Crédit d'Impôt Développement Durable.*
25. After the work has been carried out.
26. The SDTC has cost the government €7.8 billion.
27. That is an interest-free loan with no application fees, available from all leading banks in return for tax relief from the government. Available since 1 April 2009, it can no longer be combined with an SDTC. This measure had very mitigated results, partly because it allows one to make monthly repayments on loans for works costing lesser amounts and is therefore, compared to an SDTC, less tangible than a tax rebate, both for professional sectors and for consumers (Vauglin et al., 2011).
28. For a large part, renovation works in fact relate to non-relocatable jobs.

29. For a summary of works on this issue in the fields of psychology and sociology, see, for example, Moussaoui (2008).
30. In French, '*Faisons vite, ça chauffe*'.
31. These controversies crop up every so often, as can be seen with road safety policies (Hamelin and Spenlehauer, 2008).
32. http://www.liberation.fr/taxe-carbone,99830 (accessed 7 July 2012).

REFERENCES

ADEME campaign (2008), http://www-faisons.fr/IME/pdf/fiche_ADEME_lampes_basse_consommation.pdf (accessed 20 June 2013).

Akrich, M. (1992), 'The de-scription of technical objects', in B. Wiebe and J. Law (eds), *Shaping Technology/Building Society, Studies on Sociotechnical Change*, Cambridge, MA: MIT Press, pp. 205–24.

Akrich, M., M. Callon and B. Latour (2006), *Sociologie de la traduction. Textes fondateurs*, Paris: Les Presses Mines Paris.

Badinter, E. (2010), *Le conflit: la femme et la mère*, Paris: Flammarion.

Becker, O. (1963), *Outsiders: Studies in the Sociology of Deviance*, New York: The Free Press.

Benvegnu, N. (2011), 'La politique des Netsroots, La démocratie à l'épreuve d'outils informatiques de débat public', Thèse de l'école des Mines de Paris, CSI, Paris.

Besson, D. (2008), *Consommation d'énergie: autant de dépenses en carburants qu'en énergie domestique, INSEE Première*, **1176**, February.

Blanchard, P. (2009), 'Les médias et l'agenda de l'électronucléaire en France, 1970–2000', Thèse de doctorat en Science politique, Paris Dauphine, Paris.

Borraz, O. and V. Guiraudon (eds) (2008), *Politiques publiques, Tome 1, La France dans la gouvernance européenne*, Paris: Presses de Sciences Po.

Borraz, O. and V. Guiraudon (2010), 'Introduction. Les publics des politiques', in B. Olivier and V. Guiraudon (eds), *Politiques publiques, Tome 2, Changer la société*, Paris: Presses de la Fondation Nationale des Sciences Politiques, pp. 11–27.

Boy, D., M. Brugidou, C. Halpern and P. Lascoumes (eds) (2012), *Le Grenelle de l'environnement: acteurs, discours, effets*, France: Rapport final remis au Programme Concertation, Décision et Environnement, MEDDTL, CEVIPOF and Sciences Po.

Brugidou, M. (2008), *L'opinion et ses publics, une approche pragmatiste de l'opinion publique*, Paris: Presses de Sciences Po.

Brugidou, M. (2011), *Stigmatisation et dénonciation: entre adhésion doxique et distance critique*, Strasbourg: Congrès de l'Association Française de Sciences Politiques.

Brugidou, M. and M. Moine (2010), 'Normes émergentes et stigmatisation', *JADT 2010*, Rome, June.

Calwell, C. (2010), 'Is efficient sufficient? The case for shifting our emphasis in energy specifications to progressive efficiency and sufficiency', available at http://www.eceee.org/sufficiency/eceee_Progressive_Efficiency.pdf (accessed 8 January 2012).

Catteau, R., M. Galerne, P. Garnier and P.T. Desessarts (2010), *Politique publique*

de promotion des lampes basse consommation: enjeux et difficultés, Mémoire de Master d'Action Publique, sous la direction A. de Jobert and M.-C. Zélem, Paris: Ecole des Ponts.
Christiansen, T. and E. Kirchner (eds) (2000), *Europe in Change: Committee Governance in the European Union*, Manchester: Manchester University Press.
Dobré, M. (2002), *L'écologie au quotidien. Eléments pour une théorie sociologique de la résistance ordinaire*, Paris: L'Harmattan.
Dujin, A., G. Pocquet and B. Maresca (2007), 'La maîtrise de la consommation dans le domaine de l'eau et de l'énergie', Cahier de recherche du CREDOC, **237**, 85.
Economistes Atterrés (2012), *Changer d'économie. Nos propositions pour 2012*, Paris: LLL.
Fudge, S. (2010), 'Focus groups. Integrated European report', BarEnergy project, available at http://www.barenergy.eu/uploads/media/D27_Focus_Groups_Integrated_European_Report.pdf (accessed 8 January 2013).
Garabuau-Moussaoui, I. (2011), 'Energy-related logics of action throughout the ages in France: historical milestones, stages of life and intergenerational transmissions', *Energy Efficiency*, **4** (4), 493–509.
Garabuau-Moussaoui, I., F. Bartiaux and M. Filliastre (2009), 'Entre école, famille et médias, les enfants sont-ils des acteurs de la transmission d'une attention environnementale et énergétique? Une enquête en France et en Belgique', in N.N. Burnay and A. Klein (eds), *Figures contemporaines de la transmission*, Namur, Belgium: Presses universitaires de Namur, pp. 105–20.
Geels, F.W. (2002), 'Technological transitions as evolutionary reconfiguration processes: a multi-level perspective and a case-study', *Research Policy*, **31** (8–9), 1257–74.
Gossart, C. (2010), 'Quand les technologies vertes poussent à la consommation', *Le Monde Diplomatique*, June.
Grossman, E. and S. Saurugger (2006), 'Les groupes d'intérêt au secours de la démocratie?', *Revue française de Science politique*, **56** (2), 299–321.
Halkier, B. (2001), 'Routinisation or reflexivity? Consumers and normative claims for environmental consideration', in J. Groncow and A. Warde (eds), *Ordinary Consumption*, London: Routledge, pp. 25–44.
Hamelin, F. and V. Spenlehauer (2008), 'L'action publique de sécurité routière en France. Entre rêve et réalisme', *Revue Réseaux*, **147**, 49–86.
Hecht, G. (2004), *Le rayonnement de la France. Énergie nucléaire et identité nationale après la Seconde Guerre mondiale*, Paris: La Découverte.
Hirschman, A.O. (1970), *Exit, Voice, and Loyalty: Responses to Decline in Firms, Organizations, and States*, Cambridge, MA: Harvard University Press.
Jobert, B. (1999), 'Des États en interaction', L'Année de la régulation, 3, 77–9, English translation available in A. Menon and V. Wright (eds) 2001), *From the Nation State to Europe*.
Lascoumes, P. (2008), 'Les politiques environnementales', in B. Olivier and V. Guiraudon (eds), *Politiques publiques, Tome 1: La France dans la gouvernance européenne*, Paris: Presses de la Fondation Nationale des Sciences Politiques, pp. 29–67.
Lassègue, J. (2010), 'Pour une anthropologie sémiotique; recherche sur le concept de forme symbolique', Mémoire se soutenance d'habilitation de recherche, Paris IV.
Latouche, S. (2006), *Le pari de la décroissance*, Paris: Fayard.

Leray, T. and B. de La Roncière (2003), *Trente ans de maîtrise de l'énergie*, Arcueil: ATEE.
Meuric, L. (2006), 'L'évolution annuelle de l'énergie en France depuis 1973', *Réalités industrielles*, Ministère de l'Economie, des Finances et de l'Industrie, August, 59–61.
Moussaoui, I. (2007), 'De la société de consommation à la société de modération. Ce que les Français disent, pensent et font en matière de maîtrise de l'énergie', *Les Annales de la Recherche Urbaine*, **103**, September, 114–21.
Moussaoui, I. (2008), 'Domestic energy use: main barriers and drivers towards an energy-related behavioural change', BarEnergy report, available at http://www.barenergy.eu/ (accessed 8 January 2012).
Moussaoui, I. and O. Vaubourg (2004), *Maîtriser son confort, rendre confortable sa maîtrise: gestion des flux domestiques, maîtrise de l'énergie et valeurs de modération, chez les familles propriétaires*, rapport EDF R&D, multig.
Pautard, E. (2009), 'Vers la sobriété électrique, politiques de maîtrise des consommations et pratiques domestiques', thèse de sociologie, Université Toulouse II, Le Mirail Toulouse, France.
Rochefort, R. (2007), *Le bon consommateur et le mauvais citoyen*, Paris: Editions Odile Jacob.
Roy, A. (2006), 'L'environnement, de plus en plus intégré dans les gestes et les attitudes des Français', *Le 4 Pages IFEN*, **109**, January–February.
Rumpala, Y. (2009), 'La consommation durable comme nouvelle phase d'une gouvernementalisation de la consommation', *Revue française de science politique*, **59**, 967–96.
Saurugger, S. (2003), 'Les groupes d'intérêts entre démocratie associative et mécanismes de contrôle', *Raisons politiques*, **10**, 151–69.
Shove, E. (2003), 'Converging conventions of comfort, cleanliness and convenience', *Journal of Consumer Policy*, **26** (4), 395–418.
Shove, E. and A. Warde (1998), *Inconspicuous Consumption: The Sociology of Consumption and the Environment*, Lancaster: Department of Sociology, Lancaster University.
Stone, D. (1989), 'Causal stories and the formation of policy agendas', *Political Science Quarterly*, **104** (2), 281–300.
Vauglin, F., V. Beillan and A. Jobert (2011), *Le crédit d'impôt développement durable et l'éco-prêt à taux zéro: impacts d'instruments de politiques publiques visant à l'efficacité énergétique dans le bâtiment*, rapport EDF R&D, multig.
Wallenborn, G. and J. Dozzi (2007), 'Du point de vue environnemental, ne vaut-il pas mieux être pauvre et mal informé que riche et conscientisé?', in P. Cornut, T. Bauler and E. Zaccaï (eds), *Environnement et inégalités sociales*, Brussels, Editions de l'Université de Bruxelles, pp. 47–59.
Warde, A. (2005), 'Consumption and theories of practice', *Journal of Consumer Culture*, **5**, 131–53
Zélem, M.C. (2010), *Politiques de maîtrise de la demande d'énergie et résistances an changement. Une approche socio-anthropologique*, Paris: L'Harmattan.

9. Decoupling environmental impact from economic growth in Norway: viable policy or techno-optimistic fantasy?

Pål Strandbakken and Eivind Stø

INTRODUCTION

The debate over the environmental challenges has always contained the disagreements between so called techno optimists and the more traditional 'environmentalists', for lack of a better word (Strandbakken, 2009; Weissäcker et al, 1997). In the short hand version, this is the view that humankind's ingenuity will save the day, that a 'technological fix' will appear that will take care of problems like global warming through carbon capture, unlimited supply of clean energy and so on, versus the view that ascetic lifestyles, frugality and environmentally motivated poverty will be necessary. Today, this debate is often enacted as disagreement over the potential of decoupling. The United Nations Environment Programme's (UNEP) International Resource Panel addresses this theme in the report *Decoupling Natural Resource Use and Environmental Impacts from Economic Growth* (UNEP, 2011), but rather than advocating poverty and ascetic lifestyles, the panel focuses on decoupling; 'reducing the amount of resources such as water or fossil fuels to produce economic growth and delinking economic development from environmental deterioration' (p. 4). Without spending too much time and space on definitions, we just state that the idea of decoupling is to split negative environmental effects from economic growth.

Jackson (2009) distinguishes between relative and absolute decoupling. His main conclusion is that even if some progress has been made in reducing the ecological load (mainly the energy intensity) of growth, there is an 'absence of absolute decoupling' (p. 75). Jackson's principal message is that decoupling is necessary, but that it probably will not be enough to rise to the challenge of the urgent climate crisis. Further, he asserts that without confronting the structure of market economies it will not be possible to achieve deep cuts in resource use and emission levels. Jackson's scepticism seems to cover, or update, the old 'environmentalist' position, while UNEP is closer to the techno-optimist views.

At present, Norway is a very rich country. Its economy has been growing steadily since the early 1990s, both measured in gross domestic product (GDP) and as private affluence. The country's economy even seems to have remained largely untouched by the current global financial crisis. In this context, any signs of a reduced environmental impact from Norwegian households/consumers are potentially interesting because this would suggest that it might be possible to break the relationship between affluence and eco impact. This is the starting point of this chapter.

We have observed interesting trends in two consumption areas in Norway; trends we hold to be relevant to the debate over decoupling: firstly, and by far the most important, the average consumption of domestic energy more or less stopped growing in the early 1990s, peaking in 1998, and then tailing off; secondly, two ecologically rather positive tendencies in Norwegian meat consumption. One is that average meat consumption does not seem to have grown since 2007. This is potentially important because the greenhouse gas load of meat is very high. Another is a shift from beef to poultry, a change that is held to be beneficial to the environment.

We analyse the development in Norwegian energy consumption and meat consumption in the overall perspective of decoupling. The chapter aims to critically evaluate both the importance and the robustness of these trends and, in addition, to discuss them in a context of decoupling, also considering possible rebound effects and their implications for debates around the sociology of consumption.

DECOUPLING UNDER PRESENT POLITICAL/ ECONOMIC CONDITIONS?

Does economic growth necessarily lead to increased greenhouse gas (GHG) emissions? Intellectually, our starting point for this proposition is the previously mentioned idea that present day debates over absolute and relative decoupling to a large degree are updated versions of older debates over sustainable development versus 'counter modernity' theories, or between belief in technologically based solutions to the environmental challenge versus solutions more based on ascetic lifestyles, 'willed poverty' and radical redistribution. One way of rephrasing the question is to ask, openly and without prejudice, how far have we been able to go with primarily technically based policies and measures in efforts to realize a more sustainable society?

Theoretically, the whole world could run on solar and other renewable energy sources, but such a shift remains improbable, given the highly political nature of reconciling sustainability with real-world practicalities. Therefore, it is perhaps more realistic to ask: what types of relevant empirical evidence exist that might support the possibility of a radical decoupling under present political-economic conditions, and what kinds of counter evidence might we have to consider? When, for instance, we find that energy consumption in the household sector in Norway has decreased within a certain timescale, we will probably not be able to isolate the effect of technological measures, like separating the impact of the introduction of heat pumps from the effect of changing consumer behaviour (for example, reducing indoor temperature through turn down/switch off). We will, however, control for variables such as population size and average size of households, and we will be aware of the influence of general economic trends.

This means that we will be able to observe the occurrence and/or non-occurrence of decoupling for certain sectors and for certain geographical areas, even if we do not necessarily know the exact reasons for it.

ENVIRONMENTALISM AS ANTI-MODERNITY

In the late 1960s and early 1970s, environmental concerns in the developed world tended to be based around critiques of industrial society, of business and of technology. There appeared to be consensus that the environmental challenge was something that capitalist society would not be adaptable to, and that answers to the challenge would be found in total system change or revolution. From texts around this time, such as 'A blueprint for survival' (Goldsmith et al., 1972) and *Small is Beautiful* (Schumacher, 1973), Gert Spaargaren later identified the 'ecotopia' of this period as 'a society consisting of numerous small scale units, where people live their lives close to nature and to each other, where technology was of the proper scale' (Spaargaren, 1997, p.9).

Therefore, a society based on decentralization, participation, low-tech and ascetic lifestyles was the proposed solution to the environmental challenge from this generation of environmental commentators. At times, they achieved a certain visibility in the media but, in general, environmentalism was a fringe activity going on at the outskirts of society and quite removed from 'real' politics. There were some anti-pollution measures taken by governments and some legislation on problems with mercury and DDT (the insecticide dichlorodiphenyltrichloroethane), but generally there was little regular political activity on the environmental issues, and the environmental movement had little political influence. Neither in business circles nor in the environmental non-governmental organizations (NGOs) were there much interest in pragmatic solutions, policy instruments and compromise. This, however, seemed to change in the 1980s.

SUSTAINABLE DEVELOPMENT AND ECOLOGICAL MODERNIZATION

When the World Commission on Environment and Development's report *Our Common Future* (WCED, 1987) introduced the concept of 'sustainable development' in the 1980s, the above ideas from the 1970s were revisited, but in a conscious attempt at breaking with some key assumptions that lay behind those debates. Of particular importance was the re-evaluation of economic growth and of what Spaargaren (1997) has described as the politics of 'counter modernity'. While the first wave of environmental debate, following the Stockholm Conference of 1972, focused on 'limits to growth' (Meadows et al., 1972) and 'zero growth', the Brundtland Report initiated a debate over the content of growth. This is a discussion that we, a quarter of a century later, are still engaged with; a discussion that is often termed as a debate over the 'decoupling' of environmental impact from economic growth (Jackson, 2009).

The roots of the thinking behind *Our Common Future* can be traced to discussions in Germany, where Huber (1982) and Jänicke (1986) initiated a renewal of the theorizing of relations between society and the environment; a tradition that later became known as 'ecological modernization' (Hajer, 1995; Weale, 1992). Ecological modernization perspectives brought with them a critique of counter modernity and counter productivity thinking. This suggested a paradigmatic change. From then on, the ruling idea was that the sustainable society would be one in which growth, innovation and development would play a central role. We shall not go deep into the history of the more recent link between social theory and the environment here, just provide a summary of some of the important points where ecological modernization distanced itself with the environmentalism of the 1970s:

- Contrary to the more apocalyptical debates in the 1970s, ecological modernization now argued that environmental problems could be solvable inside the existing political and economic system.
- Ecological modernization introduced a more positive view of the role of technology, innovation and science, without falling back on the narrowly defined pathway of a 'technological fix'.[1]

- Ecological modernization introduced the idea that ecological investments might actually be 'profitable', an idea completely at odds with the dominant thinking of the 1970s.
- There was now a vision around a more active role for political authorities cooperating with the more progressive actors in business, in addition to involving business itself in regulation and legislation. Advocates of ecological modernization felt that this would encourage more 'incremental change' rather than an abrupt rupture.
- There was now a growing realization that a lot of environmental problems are transnational, encouraging international agreement and control regimes.

Due to subtle changes in the language of environmentalism, ecological modernization today might actually mean something like 'belief in a technological fix', whereas the positive view of technological contributions was only one aspect of ecological modernization in the 1980s and 1990s (Orsato and Clegg, 2005; York and Rosa, 2003).

To a large degree, the call for sustainable development, in the World Commission report seemed to have adopted these perspectives in a combined approach. On the environmental issue, the most striking difference between the two perspectives is the different geographical approach to the debate, where the main focus of ecological modernization seems to be The Organisation for Economic Co-operation and Development (OECD) with sustainable development covering the United Nations. The latter approach appears to be more concerned with world poverty, the North-South gap, and the environmental problems in the developing world. This 'theoretical split' implies that, in the present context, the debate over decoupling is framed by the debate around ecological modernization. Ecological modernization is about solutions for modern, relatively rich and technologically advanced societies with large populations. Modern proponents of this positive, optimistic and solution-oriented perspective include the Wuppertal Institute, famous for the 'Factor Four' perspective (Weizsäcker et al., 1997), where they argued that it would be possible to increase the global wealth substantially, and reduce the global environmental impact simultaneously.

WHAT IS DECOUPLING?

Jackson (2009) has drawn a distinction between absolute and relative decoupling[2]:

> Relative decoupling refers to a decline in the ecological intensity per unit of economic output. In this situation, resource impacts decline relative to the GDP. But they don't necessarily decline in absolute terms. Impacts may still increase, but at a slower pace than growth in the GDP. (p. 67)

His conclusion is that even if some progress has been made in reducing the ecological load (mainly the energy intensity) of growth, there is an 'absence of absolute decoupling' (p. 75). Jackson's main message is that decoupling is necessary, but that it is not likely that technological measures alone will make the necessary change; particularly with regard to the scale of the environmental problems. Other reports, such as the above-mentioned UNEP report (2011), are slightly more optimistic than Jackson's with regard to the scope that decoupling offers for enabling progress towards a more sustainable, low-carbon society. The disagreements might be due to different perspectives; 'decoupling optimists' tend to argue from what is possible, while Jackson's arguments are more empirical; his focus is on what have been accomplished so far. However, both sides agree that decoupling is taking place.

The main reason why we have not seen evidence of absolute decoupling so far is that economies have grown faster than the technology revolution has been able to deal with by way of efficiency. Another reason relates to the 'x-factor' of the rebound effect, as Sorrell and Herring have pointed out:

> To achieve reductions in carbon emissions, most governments are seeking ways to improve energy efficiency throughout the economy. It is generally assumed that such improvements will reduce overall energy consumption, at least compared to a scenario in which such improvements are not made. But for many years, economists have recognized that a range of mechanisms, commonly grouped under the heading of rebound effects, may reduce the size of the 'energy savings' achieved. (Sorrell and Herring, 2009, p. 3)

According to Sorrell and Herring (2009, p. 6), an example or illustration of this might be more fuel efficient cars. A fuel-efficient car will lower your petrol bills, so that you might afford to take your family on a holiday to Spain, by air. Or lower running costs might encourage you to drive longer distances and/or more frequently. Throne-Holst et al. (2007) argue, however, that even when acknowledging the rebound effect, consumer choice matters. Consumption is not necessarily an environmental zero-sum game. And the anticipated rebound effects should never be an argument against efficiency improvements.

The concept of the 'environmental Kuznets curve' is a useful way through which to explore this discussion (Stern, 2003). The environmental

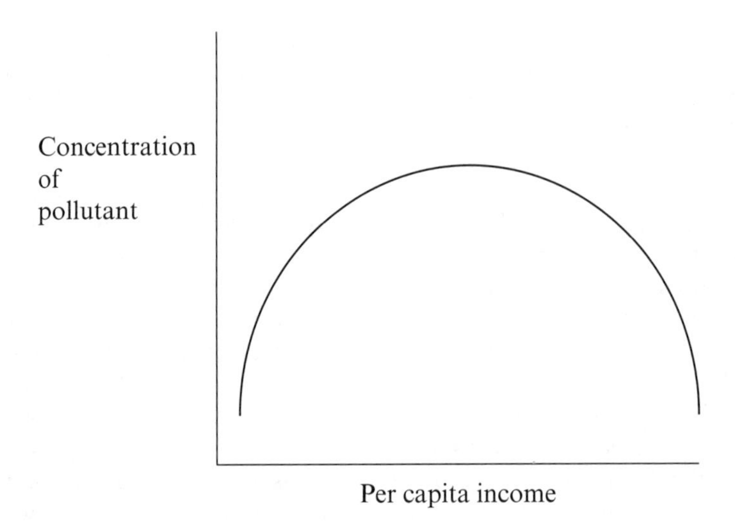

Figure 9.1 The environmental Kuznets curve

Kuznets curve (Figure 9.1) is an inverted, U-shape curve, demonstrating the relationship between economic growth and environmental impact for a country or a sector. During the industrial revolution, the curve was rising, meaning that the negative environmental impact grew faster than the economy. According to this theory, we have seen a shift in this relationship in the last decade or so. In a number of countries, there is evidence to suggest that economic growth is larger than the environmental impact, where observers often cite reduced air pollution and increased eco-efficiency in the industry as examples (Dasgupta et al., 2002).

The environmental Kuznets curve has been criticized, however, for a number of reasons. For example, there is an ongoing discussion about the shape of the curve (Harbaugh et al., 2000). The estimated pollution–income relationship could very well turn into an inverted S-shape, meaning that the downward bend might disappear in the future. Furthermore, it is not clear if it is relevant for all countries or relevant for all sectors of the economy. In addition, it seems problematic that the Kuznets curve theory has no concepts for the difference between absolute and relative decoupling. This is problematic because then even if we observe a reduced environmental impact, that is, in the form of reduced GHG emissions from a rich European nation, we would not know how much of the reduction that really comes from technical improvements in industrial production and how much that comes from industry being shifted to Southeast Asia.[3]

A second influential concept which deals with the relationship between absolute and relative decoupling is the 'arithmetic of growth', developed by Erhlich in the 1960s (Erhlich, 1968). The impact of human activity I is the dependent of three crucial factors: P (the size of population), A (affluence) and T (the technology factor). This relationship is presented in the following way:

$$I = P \times A \times T$$

In this model, the definition of relative decoupling might be illustrated as a reduction of T; technological improvement or improved efficiency. Absolute decoupling, on the other hand, would demand a reduction in I: total environmental impact (Jackson, 2009 p. 77). Population size will (hopefully) not vary because of short-term environmental policies,[4] and reduced economic growth has not been targeted as desirable by any government to date. This leaves technology – in a wide sense where it also includes consumer practices and interactions with the technologies – as the only variable factor in the equation.

ABSOLUTE OR RELATIVE DECOUPLING?

There has been a substantial growth in the world population as well as in general affluence during the last few decades. For absolute decoupling to take place, this would mean that increased technological efficiency would have to match the combination of increased population size and growing affluence. In the future, we would need to decouple economic activity from the environmental impact of this activity, and we would need to move from relative to absolute decoupling. In light of this argument, Jackson presents a case for the significance of absolute decoupling. For addressing the challenge of climate change – and also the limits on resources – such a move would seem to be fundamental.

However, we might argue that Jackson's insistence on the necessity of absolute decoupling could be a potential constraint to the development of other pathways to sustainability. On the level of global emissions and impact, he is right; increasing populations with increasing affluence have made an impact that improved technologies have so far not been able to match. It is, however (as Jackson notes), also important to point out that without technologies and practices for (relative) decoupling, we would have been much worse off.

The problem, as we see it, is that to approach all policies from a global perspective could be very discouraging. Politicians, citizens, consumers, researchers and activists would find it meaningless to engage in politics for environmental change if decoupling/environmental achievements could not be celebrated for smaller units of progress than the biosphere. For instance, if a large city, like Berlin or London, should be able to significantly cut its GHG emissions, it serves no useful purpose to have to point out that because of Shanghai's growth nothing has been gained in absolute terms. It is interesting to observe and to celebrate examples of absolute decoupling of environmental impact from economic growth in nations, regions and municipalities, and even in sectors of the economy.

There are two reasons why this should be the case. First, the developed parts of the world still have a higher per capita environmental impact than the new and expanding economies in Asia, South America and Africa. As long as this is so, the moral position of the West will remain fragile. The key actors in the global climate negotiations are China and the US, and critics of the relative failure of the United Nations Framework Convention on Climate Change (UNFCCC) have pointed out that they are the ones that must agree to sharp cuts in emissions. This is probably a reasonable argument to make, but China's population is approximately three times bigger than that of the US. Surely the only fair yardstick here has to be per capita impact. If we demand reduced emissions from the Asian nations, we have to demand them from ourselves first. Second, in the way the West has been a model for the rest of the world (Üstüner and Holt, 2010). 'They' imitate 'our' lifestyles. Hence, we should supply them with a low carbon lifestyle to imitate. In this perspective, absolute decoupling in units smaller than the entire biosphere is relevant, interesting and important.

A SURPRISE SUCCESS? THE CASE OF NORWAY

As mentioned, Norway today is one of the richest countries in the world, with a constantly growing economy since the early 1990s, as measured in GDP or as private affluence. The Norwegian economy was largely untouched by the current global financial crisis, even though the growth in private consumption slowed down to 1.6 per cent in 2008 and 0.2 per cent in 2009 (Hille, 2011). In a situation like this, any signs of a reduced environmental impact from Norwegian households/consumers are potentially interesting, because it could mean that the relationship between affluence and impact on the environment might be broken. The 'case of Norway' is rather different from the situation in the rest of Europe, with approximately 100 per cent renewable energy supply from hydropower, a rather small population in a very large country as well as huge exports of natural gas and oil. Nevertheless, there are some trends that could provide interesting insights into the arguments proposed by Jackson, particularly

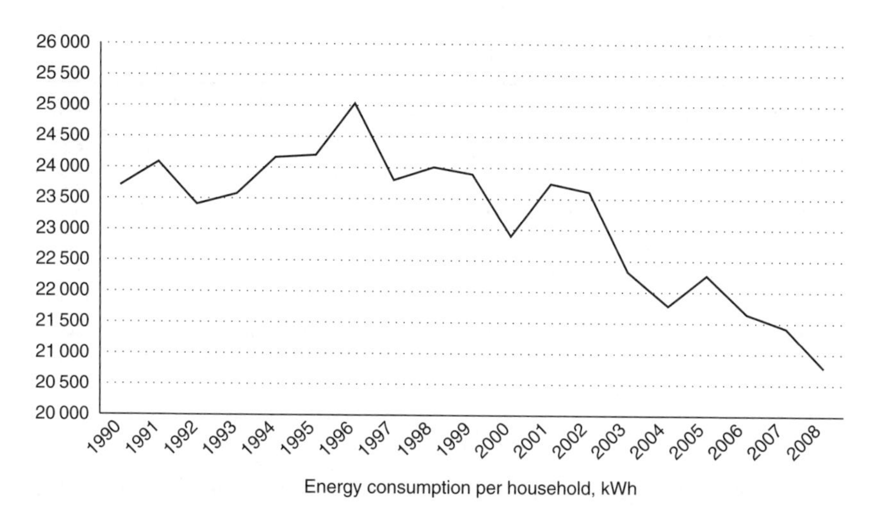

Figure 9.2 Average energy consumption per household 1990–2008, Norway

in terms of what can happen with environmental impacts when certain levels of affluence are reached in a population.

We have observed developments that probably are relevant to the debates on decoupling. The average use of domestic energy more or less stopped growing in the early 1990s, peaking in 1998 or a bit earlier, and since then it has decreased. Trends in Norwegian meat consumption point in the same direction. Total meat consumption seems to have peaked around 2007. This is interesting because the GHG load of meat is very high. If this is right, it has stabilized on a significantly lower level than (less rich) neighbouring countries. In addition, the whole increase in total per capita meat consumption from 1997 to 2007 is explained by poultry, which is far less problematic than beef with regard to carbon emissions.

Our initial observation revolved around the changes in energy consumption in Norwegian households. In the project 'A secret success' (Stø et al., 2011),[5] we documented the appearance of a real decrease at the household level in energy consumption in general, but also specifically for electricity. Figure 9.2 shows the average energy consumption per household for the last two decades. There is also evidence for reduced energy use at the individual level. The energy consumption per person has reached its peak, and is declining.

Even with a slight increase in 2009, mainly due to very low temperatures in the winter season, the general trend remains in decline. Energy use per person was the same in 2009 as it was in 1990, and 9 per cent lower than in

Table 9.1 Meat consumption (kg) in Norway 1979–2009

Type of meat	1979	1989	1996	1997	1998	2007	2008	2009
Pork	19.5	17.7	23.7	24.0	23.8	26.9	26.0	25.9
Beef	18.5	16.8	20.1	19.9	20.5	20.0	20.1	18.5
Veal	0.5	0.3	0.2	0.2	0.2	0.4	0.4	0.4
Lamb	5.5	6.0	5.2	5.6	5.5	5.6	6.0	5.2
Poultry	2.7	4.6	6.9	7.4	7.7	14.9	16.9	17.7
Others	0.3	0.3	0.2	0.2	0.2	0.2	0.1	0.1
Total	47.0	45.7	56.3	57.3	57.9	68.0	69.5	67.8

Source: Svennerud and Stein (2011).

1995, when the long-term upward trend in this area stopped (Hille, 2011, p. 34).

Norway's population increased by 14 per cent between 1990 and 2009, and this was also a period of economic growth. The yearly growth in private consumption per inhabitant was 2.6 per cent from 1990 and to the present (Hille, 2011, p. 12). This should mean that, seen in the perspective of Erhlich's arithmetic of growth, we have witnessed a reduction in *I* (impact), even with a strong growth in *A* (affluence) and a moderate growth in *P* (population size) since 1995. If we isolate the country (Norway), and the sector (domestic energy use), we should be able to present this as an example of absolute decoupling: reduced environmental impact in a period of economic growth.

We did observe parallel tendencies in other consumption areas, particularly this promising trend.

What does available data tell us about the development of Norwegian meat consumption? Data from the Norwegian Agricultural Economic Research Institute (NILF) and the farmers' cooperatives are presented in Table 9.1. Two trends are worth highlighting. Firstly, Norwegian meat consumption is lower than in the neighbouring countries. In 2005, Denmark consumed 90 kg, Sweden 79 kg, Finland 72 kg and Iceland 80 kg per person. From a decoupling perspective, this is interesting because it breaks the often usual trend towards increased income and increased meat eating. Norway is richer than its neighbours, but meat consumption is lower, and has remained so.

Secondly, the increase during these years is mainly a result of an increase in poultry consumption, from 2.7 kg per person in 1979 to 17.7 kg in 2009. Beef and lamb consumption have remained relatively stable. This means that the environmental impact from the observed growth is less than the total figure may indicate because poultry has smaller carbon dioxide impact than beef. This could be an extra indication that lifestyle changes in more affluent societies will not necessarily lead to increased environmental impact.

Total meat consumption has been relatively stable for the last five years. Even if we do not know if this means that consumption has stabilized on a moderate level, or if it is only a temporary pause in an upward trend, it is interesting to note that the richest country in the region still has the lowest meat consumption.

CONCLUSIONS

In this chapter, we argue that if we want to discuss if, how and to what extent consumer affluence can be decoupled from negative environmental impact, it will be beneficial to consider smaller units of analysis; particularly sectors, countries, perhaps even municipalities. This contrasts with Jackson's (2009) analysis, which seems to define decoupling in a way that makes it potentially rather unfruitful for political debate. While there are obvious environmental benefits to be gained by absolute decoupling, there is evidence to suggest that our 'relative' decoupling can contribute to reducing environmental impacts. In fact, these smaller units might be regarded as 'social laboratories' and niche activities from where positive experiences could be scaled up to even greater effectiveness.

For both domestic energy use and meat consumption – two environmentally and economically important consumption areas – the research charted in this chapter reveals some interesting trends; both relevant for the debate on decoupling.

The chapter has highlighted the fact that average Norwegian consumption of domestic energy more or less stopped growing in the early 1990s, peaking in 1998, and since then decreased. We believe that this is mainly explained by two phenomena. Firstly, it seems as if we have reached some kind of saturation point. When indoor temperatures reach 23°C, there is for most people little or no added value in further increases. Something similar might have happened with dwelling sizes. With approximately 55 square metres per person (Hille, 2011, p. 30), there is probably less need for further increase. The theory, then, is that when indoor temperatures and household sizes have stabilized, the technology factor starts to take effect. Then heat pumps, better insulation, energy-saving light bulbs and energy-efficient appliances do contribute to reductions and override the 'affluence factor'.

Developments in Norwegian meat consumption are more uncertain, but again show some promising trends. Meat is environmentally important, because the GHG load of certain types of meat is very high. We have observed that average meat consumption does not seem to have grown since 2007. At present, we do not know why this is so, but the increased preference for poultry is consistent with general trends in Europe and the USA; where people are increasingly consuming white meat mainly for health reasons. Even if we cannot at present explain why Norwegian meat consumption has stabilized, we regard it as an interesting and promising development. We do not claim that this proves anything, but we see it as an indication that meat consumption is not completely determined by consumer affluence, which might be the present common opinion.

The overall conclusion is obviously not that Jackson is wrong. Quite, the contrary, he has defined the limits and pointed to present constraints in relation to the economics of a finite planet. However, as this chapter has argued, the concepts of relative and absolute decoupling and the relation between them might be more complex than Jackson suggests. We argue that relative decoupling can be used to highlight environmental impacts in smaller units; both to demonstrate possibilities and to encourage more or less isolated experiments and developments in the direction of sustainable consumption.

NOTES

1. The idea of a 'technological fix' is based on the assumption that due to mankind's ingenuity, technical solutions to our environmental problems can be produced, like solving climate problems with carbon capture and new sources of energy and more efficient use of energy, that materials technology will be able to substitute any scarce raw material and so on.
2. The UNEP (2011, p. xii) report distinguishes between 'resource decoupling' and 'impact decoupling', where the first means to use resources more effectively and the second means to reduce the total environmental impact of human activity. It seems as if impact decoupling is a concept very similar to Jackson's 'absolute' decoupling, while resource decoupling appears to be more like a tool or a method.
3. Obviously, to move manufacture out of Europe and the USA will not solve any global emission problems, unless Asian manufacturers should happen to perform environmentally better than European or American.
4. But population size is subject to long-term demographic, cultural and technological-political change; like the one child per family policy in China, removal of foetuses with unwanted gender and even reports of killings of girl babies in some countries, making populations less fertile in the long run, in addition to the often observed reduction in childbirths after societies reach a certain level of affluence.
5. The success is regarded as a secret because it has not, or only to a very limited degree, been communicated to the public, and it has not been highlighted in national media. When asked, our focus group participants all believed that domestic energy use in Norwegian households was increasing, even increasing rapidly. Why such good news is not publicized we do not know, but a fair guess is that those who are aware of the trend are the same

actors that are engaged in achieving reductions. There might be a fear that good news will undermine the work undertaken to change people's consciousness and habits.

REFERENCES

Dasgupta, S., B. Laplante, H. Wang and D. Wheeler (2002), 'Confronting the environmental Kuznets curve', *Journal of Economic Perspectives*, **16** (1), 147–68.

Ehrlich, P. (1968), *The Population Bomb*, New York: Buccaneer Books.

Goldsmith, E., R. Allen, M. Allaby, J. Davoll and S. Lawrence (1972), 'A blueprint for survival', *The Ecologist*, **2** (1), 1–44.

Hajer, M.A. (1995), *The Politics of Environmental Discourse: Ecological Modernization and the Policy Process*, Oxford: Clarendon Press.

Harbaugh, W., A. Levinson and D. Wilson (2000), 'Re-examining the empirical evidence for an environmental Kuznets curve', Department of Economics, Georgetown University, Oregon, available at http://www8.georgetown.edu/departments/economics/pdf/007.pdf (accessed 8 January 2013).

Hille, J. (2011), *Økologisk Utsyn 2011. Økologiske konsekvenser av den norske økonomiske utviklingen i året som gikk. Forbruket*, Oslo: Framtiden i Våre Hender.

Huber, J. (1982), *Die verlorende Unschuld der Ökologie. NeueTechnologien und superindustrielle Entwicklung*, Franfurt am Main, Germany: S. Fischer.

Jackson, T. (2009), *Prosperity without Growth: Economics for a Finite Planet*, London: Earthscan.

Jänicke, M. (1986), *Staatsversagen. Die Ohnmacht der Politik in der Industriegesellschaft*, Munich, Germany: Piper.

Meadows, D., J. Randers and W.W. Behrens (1972), *Limits to Growth, A Report to the Club of Rome*, New York: Universe Books.

Orsato, R.J. and S.R. Clegg (2005), 'Radical reformism: towards critical ecological modernization', *Sustainable Development*, **13**, 253–67.

Schumacher, E.F. (1973), *Small is Beautiful*, London: Blond and Briggs.

Sorrell, S. and H. Herring (2009), 'Introduction', in H. Herring and S. Sorrell (eds), *Energy Efficiency and Sustainable Consumption: The Rebound Effect*, New York: Palgrave Macmillan.

Spaargaren, G. (1997), *The Ecological Modernization of Production and Consumption. Essays in Environmental Sociology*, Wageningen the Netherlands: Landbouw Universiteit Wageningen.

Stern, D.I. (2003), 'The environmental Kuznets curve', *Internet Encyclopaedia of Ecological Economics*, Troy, NY: Department of Economics, Rensselaer Polytechnic Institute.

Stø, E., N. Heidenstrøm, P. Strandbakken and H. Throne-Holst (2011), 'The secret success of the reduction in the Norwegian electricity consumption', paper presented at 'Energy Efficiency First: The Foundation of a Low-Carbon Society, ECEEE 2011 Summer Study, Hyems, France 6–11 June.

Strandbakken, P. (2009), 'Energy fools the technician. Product durability and social constraints to eco efficiency for refrigerators and freezers', *International Journal of Consumer Studies*, **33** (2), 146–50.

Svennerud, M. and G. Steire (2011), *Beregning av det nonske Kjøttforbruket*, Oslo: Norwegian Agricultural Economic Research Institute (NILF).

Throne-Holst, H., E. Stø and P. Strandbakken (2007), 'The role of consumption and consumers in zero emission strategies', *Journal of Cleaner Production*, **15**, 1328–36.

UNEP (2011), *Decoupling Natural Resource Use and Environmental Impacts from Economic Growth*, Geneva: United Nations Environment Programme.

Üstüner, T. and D.B. Holt (2010), 'Toward a theory of status consumption in less industrialized countries', *Journal of Consumer Research*, **37**, June, 37–56.

Weale, A. (1992), *The New Politics of Pollution*, Manchester: Manchester University Press.

WCED (World Commission on Environment and Development) (1987), *Our Common Future*, Oxford: Oxford University Press.

Weizsäcker, E., A.B. Lovins and H. Lovins (1997), *Factor Four: Doubling Wealth, Halving Resource Use*, New report to the Club of Rome, London: Earthscan.

York, R. and E.A. Rosa (2003), 'Key challenges to ecological modernization theory', *Organization and Environment*, **16** (3), 273–88.

PART III

Different policy approaches from an international perspective

10. Living smart in Australian households: sustainability coaching as an effective large-scale behaviour change strategy

Colin Ashton-Graham and Peter Newman

INTRODUCTION

This case study presents the methods, results and learnings from an innovative and large-scale behaviour change programme addressing, in a holistic manner, the major contributors to the carbon footprint of households.

The Living Smart Households project was developed by the Department of Transport in Western Australia to build upon successful behavioural interventions in small group sustainability and large-scale transport and water demand management. Two large-scale demonstration projects have been completed, across 25 000 households, testing different strategies and techniques both within and between projects. Each project had a budget of around $AUS 2 million and was delivered (research, service delivery and evaluation) over a two-year period.

The Living Smart Households project provides a best practice example of programme design that is based on research insights, connected to formative research in the target community and monitored and evaluated across process, output and outcome measures.

The successes of the Living Smart Households project include strong market penetration (engaging with around 60 per cent of target households), large-scale operation (managed across more than 10 000 households), significant behaviour change (households adopting conservation behaviours such as short showers, switching off standby power loads and installing solar electric panels) and independently verified outcomes (around a 6 per cent reduction in energy, water and car use across programme participants compared to control). The benefits of a programme that addresses many aspects of household sustainability include a coordinated service to households as opposed to agencies competing for the attention of customers and a cost-effective outcome secured by utilizing a single engagement process to produce multiple outcomes.

The learnings are that a facilitative or coaching approach is more powerful than an approach based upon service provision. The second Living Smart demonstration programme achieved an 8 per cent reduction in electricity use across a participant group that had low socio-economic status and hence was hard to reach. Within this facilitative framework it is demonstrated that a process of self-framing allows individuals with different environmental, monetary and practical attitudes to each adopt pro-environmental behaviours. It is also shown that commonly held values, including the importance of water conservation to Western Australians, can be utilized through the coaching practice of engaging households in expressing their values, that is, self-framing, to trigger behaviour change in unrelated areas such as energy, waste and travel choices.

The chapter concludes by integrating the practical learnings back into one of the leading theoretical models of behavioural determinants. This learning by doing approach has set up the potential for further improvements to an already successful behaviour change intervention.

POLICY CONTEXT

In the mid 1990s the Government of Western Australia developed an innovative policy setting to deploy behavioural travel demand management as a complement to supply-side investments (Government of Western Australia, 1995). This policy mix was in response to an unsustainable pattern of land use and transport development in Perth that had resulted in a sprawling, low density city with a level of car dependence on a par with the worst cities in the world. In 1991 car ownership was at near saturation levels of 551 cars per 1000 persons (all ages) and approximately 78 per cent of all person trips in metropolitan Perth were made, either as driver or passenger, by car. Despite the sprawling development with suburban land uses separated from centres of services and employment, the overall travel demand in Perth was not dissimilar to more sustainable cities. Between 1986 and 2000 the mean trip distance in Perth increased from 7.6 kilometres (km) to 8.0 km and the number of trips per person day increased from 3.4 to 3.5 (Socialdata, 2001). Despite short average trip distances (and a distribution of trip distances similar to most modern cities at around 10 per cent of trips below 1 km, 30 per cent below 3 km and 50 per cent below 5 km), the choice of travel mode was twice as car dependent as more sustainable cities (Newman and Kenworthy, 1999).

The drivers of change in Perth for sustained policy development to reduce car dependence can be seen politically in a series of election campaigns fought over the need for renewing the rail system from 1983 to the present day (Newman, 2012). This led to dramatic commitments to electrify, extend to the north and extend to the south, leading to a 172 km modern electric rail system. The patronage on this system went from 7 million passengers a year to 70 million a year. At each stage of the transport planning for supply-side improvements there was a parallel demand from the public for better information about transport options. This was outlined first in the 1995 Metropolitan Transport Strategy.

The Metropolitan Transport Strategy was made operational through a plan called TravelSmart, which meant to deliver a Perth Bicycle Network, expanded public transport services and a behavioural demand management programme. The Government of Western Australia appointed leading German public transport marketing consultants, Socialdata, to deliver a multi-modal marketing service for Perth households. The research observation, discussed in more detail below, was that the theoretical potential existed to transfer half of all car trips to alternative modes of travel. The Travel Smart Household programme, of multi-modal travel choice marketing, was rolled out to approximately one third of the Perth Metropolitan population (200 000 households) between 1996 and 2007 achieving around a 10 per cent reduction in car trips in the programme areas (Ashton-Graham and John, 2006).

The Travel Smart Household innovation triggered similar marketing interventions to reduce water use (Water Corporation, 2009), an analysis of the impacts of Travel Smart on physical activity (Ashton-Graham and Putland, 2007) and ultimately an Active Smart intervention to increase physical activity (Western Australia Sports Federation, 2010). In each case Socialdata was engaged as the delivery contractor and changes in behaviour of around 10 per cent were achieved. The success of these 'Smart' programmes stems from a large potential for change (that is, genuine choice of travel modes, large amounts of wasted water or plenty of discretionary time in which to exercise) and a combination of information and motivation gaps driving suboptimal pre-programme behaviours.

The Australian Government (federal) has driven innovation in carbon abatement through a series of grant programmes that partner with industry, the voluntary sector and state and local governments. A national Travel Smart behaviour change programme was funded between 2003 and 2006 under the Greenhouse Gas Abatement programme (RED 3, 2006). The Travel Smart programme delivered cost-effective carbon abatement to the federal government and built capacity across state jurisdictions and amongst an emerging group of travel behaviour change contractors.

The leading contractors also took the evidence and learnings from this Australian Government programme and secured demonstration projects in Europe and North America (Cairns, 2004).

Following a period of sustainability policy development (Newman, 2005), the Government of Western Australia launched a Premier's Action Statement on Climate Change (Government of Western Australia, 2007). This policy and programme initiative included an endorsement of the carbon abatement outcomes of the Travel Smart programme along with dedicated funding to develop and pilot a more holistic approach to managing the carbon footprint of households. The Western Australian Department of Transport engaged with leading consultants and with a range of state, local government and non-government organizations to design an intervention. The central inputs to this development phase were the Travel Smart Household programme (developed by the Western Australia Department of Transport), small group Living Smart courses (developed by the City of Fremantle, Southern Metropolitan Regional Council and Murdoch University), a community-based social marketing framework, coaching conversation techniques (Ampt and Ashton-Graham, 2008), psychology theory and logistical methodologies. The resulting Living Smart Households project was piloted in 2008–09 and further developed in 2010–11. This innovative programme, addressing the whole household carbon footprint, is the central case study presented in this chapter.

During the pilot-testing phase of Living Smart Households, the Garnaut Review of the Economics of Climate Change (in Australia) identified the personalized coaching approach of Travel Smart and Living Smart as a highly effective tool in carbon abatement. Garnaut concluded that 'basic media campaigns and pamphlets are often neither targeted nor tailored and there is considerable evidence that their effectiveness is limited . . . . Programmes need to be targeted and tailored to ensure that the right individuals receive suitable information. This seems to be done particularly well in the western Australian government's Travelsmart program' (Garnaut, 2008 p.409). Garnaut commissioned a case study on the Travel Smart and Living Smart innovation in Western Australia, concluding that the programme has the potential to deliver cost effective carbon abatement together with other socio-economic benefits (Ashton-Graham, 2008).

Most recently the Australian Government developed a Solar Cities programme to drive household energy efficiency in tandem with smart grid and solar technologies. A consortium of state, local government and private sector entities secured Perth Solar City programme funding over a four-year period commencing 2009–10. The innovation in the Perth Solar

City was to deploy the Living Smart Households project as the main community engagement and behaviour change strategy. The aims of Living Smart Households, within Perth Solar City, were to deliver household energy-efficiency gains through behaviour changes and new technology adoption across stationary energy, transport energy and the embedded energy in water use and wider consumer choices.

THE LIVING SMART APPROACH

The Living Smart Households project has been designed to extend the reach (number of persons engaged) of the innovative small group Living Smart courses that were developed by the Southern Metropolitan Regional Council, City of Fremantle and Murdoch University. The small group courses are designed to engage with approximately 30 people over seven (weekly) two-hour sessions. The course is self-directed in that the participants determine the amount of time allocated to each sustainability topic, contribute their own experience and expertise and select site visit or practical experiences to supplement the group discussion sessions. The course is a holistic sustainability facilitation that is grounded in adult learning principles and goal setting.

The range of topics covered include: Making Personal Changes (goal setting); Climate Change; Power Smart; Water Smart; Simple Living (home-made and recycled); Move Smart (travel choices); Healthy You (exercise, diet and stress); Healthy Homes (green cleaning and materials); Gardening for Productivity; Gardening for Biodiversity; and Community Smart (engaging and volunteering). Evaluations of the Living Smart courses have found significant changes in environmental knowledge and self-reported sustainable behaviours (Sheehy, 2003). Although more than 1000 people have participated in Living Smart courses since 2007, the small group and organic nature of the programme has not yet been consolidated into an outcomes-based (meter reading) evaluation.

The adaptation of the Living Smart course into a Living Smart Households project uses mail and telephone engagement to provide information, feedback and advice to households over a period of eight months including about six individual contacts (Government of Western Australia, 2012a). This approach allows thousands of households to engage with Living Smart sustainability topics, the majority of whom would not attend a small group course. It also allows a longer-term relationship between the service provider and the household and facilitates feedback on changes in metered energy and water usage. The limitations of the household approach include much less contact time (around one hour per household compared to 14 hours of course time), a smaller set of sustainability topics (because households self-select from the available suite of information) and limited opportunity for participants to interact with each other. The inclusion of small group Living Smart courses, open to Living Smart Households, participants, attempts to overcome these limitations for the small number of participants (approximately 3 per cent) who choose to participate in the course experience.

Living Smart Households incorporates the proven goal-setting techniques from Living Smart courses by deploying coaching techniques. The first pilot programme provided coaching in the form of telephone support to interpret meter reading feedback and to ask households to set targets for reducing their energy, water and car usage. Perth Solar City, the second Living Smart Households programme, deployed much more sophisticated database systems and conversational techniques to assist households to identify the gap between their attitudes and behaviours and to choose, and commit to, specific sustainability actions. The key drivers for this change in technique include allowing households to choose their own motivations (frame) for change, to identify their own barriers and benefits for specific changes (cognitive dissonance), to engage in specific agreements on actions (progress measures) and to include instructional feedback (injunctive norms). More recently Ashton-Graham has deployed feedback on ranking of water use (comparative social norms) in a separate water demand management programme. The research behind each of these behaviour change techniques is discussed further below.

BEHAVIOURAL RESEARCH

The initial insight driving the Department of Transport's involvement in behavioural demand management was research into the 'in-depth potential' for travel behaviour change (Socialdata, 2000). This research took one-day travel diaries from 820 survey respondents across five metropolitan local government areas, representative of inner, middle and outer urban areas, and performed a desk analysis of the theoretical travel alternatives available to them. Each survey respondent was then interviewed to find out the degree to which they were aware of the availability of travel alternatives, could command sufficient knowledge (for example, timetable details) to access the option, correctly perceived the relative advantages (for example, travel time) of the chosen mode (and its alternatives) and made an informed choice. The resulting mode choice, constrained trips and drivers of the choice to use the car are set out on average across all 820 respondents in Figure 10.1.

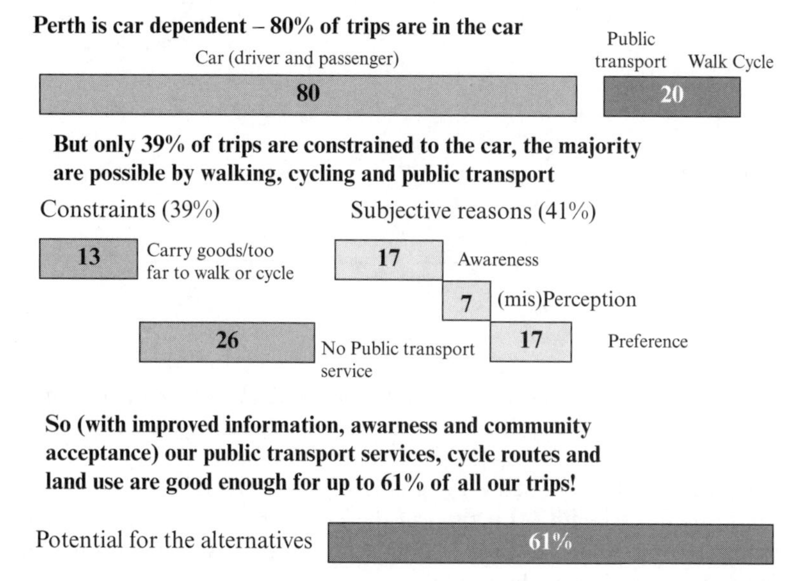

Figure 10.1 Potential to change car trips to travel alternatives (and the barriers to change)

The finding was that up to half of car trips were theoretically replaceable with an alternative travel mode and that a lack of awareness (information), misperception (of quality) and preference were all significant determinants of mode choice.

Scoping the transferability of the Travel Smart programme to energy (and water), the Department of Transport accessed depersonalized data from household meter readings. At a simple observational level it was apparent that, within neighbourhoods with similar housing stock and occupancy, electricity use per household per day ranged from 4 units (kWh) to more than 200 units per day with a mean of 16 units, mode of 9 units and 10 per cent of households using more than 29 units per day. Water use has a similar spread with the top 20 per cent of households using around 500 litres per person per day and the bottom 20 per cent using around 100 litres (Figure 10.2).

In the 1970s a detailed study of similar townhouses in the US found a spread of energy consumption of around 3:1 after adjustment for any differences in house standard and occupancy (Socolow and Sonderegger, 1976). The observation that occupant variability is a greater influence on energy and water use than housing design/standards is striking in a policy

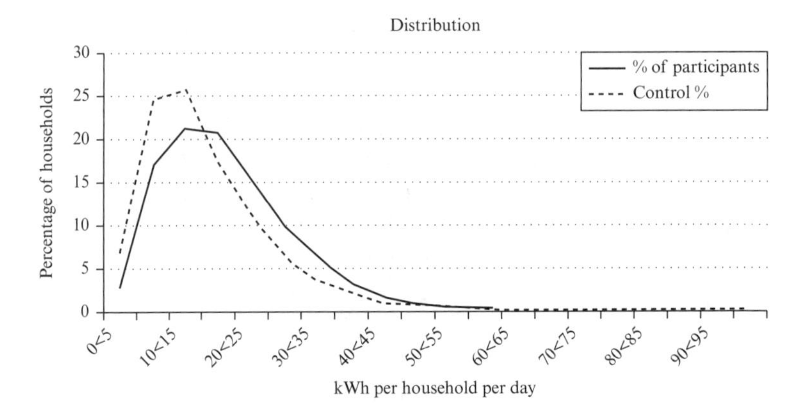

Figure 10.2 Spread of energy use (same neighbourhood) = wasted energy = behaviour

environment that is focused on engineering standards. To investigate the human processes behind this finding the Department of Transport commissioned home and face-to-face interviews on energy, water, travel and waste behaviours (Socialdata, 2007).

This research found that strong pro-environmental attitudes were stated, but few sustainability behaviours revealed. Such an observation is consistent with research into pro-environmental behaviours, which observe a gap between pro-environmental knowledge or attitudes and actual behaviours (Kollmuss and Agyeman, 2002). The in-depth interviews, taking a social marketing approach, explored the barriers to and benefits of the behavioural choices made by households.

The Department of Transport convened an Expert Group of demand management programme managers and environmental groups to explore the scope and design for a household sustainability programme. It engaged a leading community-based social marketing consultant, Doug McKenzie-Mohr (2010), to facilitate the initial workshop. What emerged from the workshop was the need for a coordinated and holistic programme and the potential to combine the powerful small group Living Smart community course with the large-scale marketing approach of Travel Smart.

The Department of Transport entered into a branding agreement with Living Smart (now a community not-for-profit foundation) and engaged a number of leading environmental communicators to develop the tools and messages in a form that could be deployed large scale. Sustainable gardening personality, Josh Byrne (2006), and sustainability author, Tanya Ha (2003), were engaged to co-author materials and act as ambassadors to the Living Smart Households programme. Behaviour change consultants were engaged to develop the coaching conversation that would enable household behaviour changes and test written communication tools.

Between 2008 and 2011 the Department of Transport managed the delivery of two Living Smart Households demonstration projects reaching a combined total of 25 000 households in Perth (Government of Western Australia, 2012b). A process of learning by doing and continuous improvement led to an evolution of the behaviour change methodology. The formative research and responses of programme participants suggested that the influences on water and energy behaviours were more values based than travel choices (information and motivation based). The Department of Transport shifted the approach from one characterized by information service provision, encouragement and feedback to one characterized by facilitation, specific agreements and feedback (discussed further in the Methodology, Monitoring and Evaluation sections below).

Further, project design has been informed by research and field experiments in psychology and behavioural economics. Of particular influence on the design of Living Smart Households have been:

- The deployment of descriptive norms, by Nolan et al. (2008), where it was observed that normative messages about the behaviours of others in the community had a stronger influence on behaviour change than messages based on facts or potential motivating factors such as environmental responsibility.
- Injunctive norms in the OPower programme, evaluated by Costa and Kahn (2010), which demonstrated a modest reduction in energy use in response to both normative comparisons of usage and injunctive commentary in the form of sad or happy faces. This evaluation also found that political persuasion was a determinant of consumer reductions or increases (reactance) in energy use.
- Theory of Planned Behaviour, by Ajzen (2006), identifies a set of behavioural determinants (such as attitudes, norms and perceived control) that can be targeted to influence behaviour.

The operational aspects of Living Smart draw upon applied research from a number of behaviour change practitioners. These include the barriers and benefits of both existing and desired behaviours identified by McKenzie-Mohr (2010), which has been estimated by the panel of experts engaged by the Department of Transport and tested in focus groups in order to identify key messages as well as the emotive frames and specific 'asks' discussed by Robinson (2010).

These research insights and the process of learning by doing has resulted in an approach that is centred on coaching engagement with households.

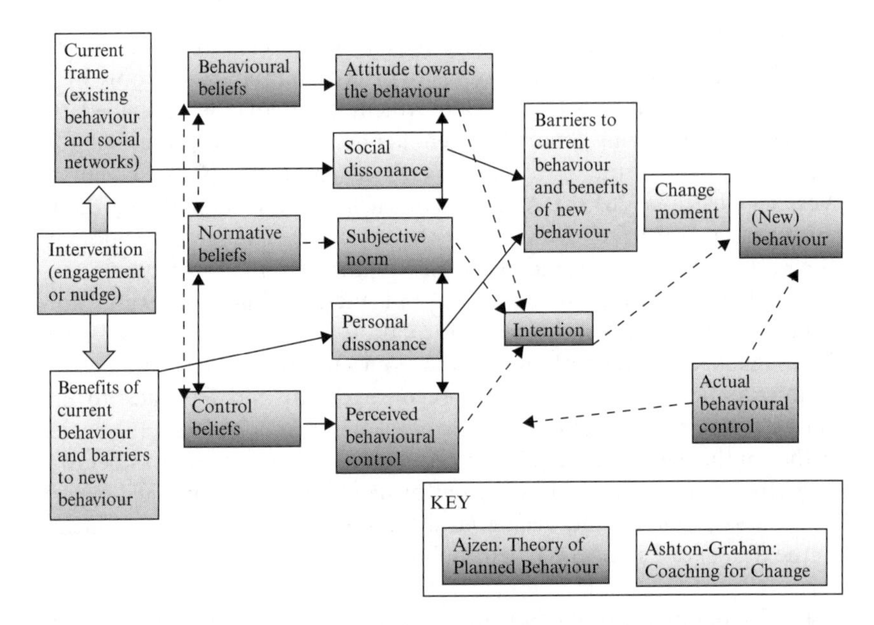

Figure 10.3 Theory of planned behaviour – applied to behaviour change

The approach, being based on research insights from Australia, US and Europe, is likely to be transferable to populations in other developed cities.

Figure 10.3 was developed by Colin Ashton-Graham to show how the coaching approach interacts with Ajzen's determinants of behaviour. In particular, by inviting the participant to express their motivations and aspirations, the coaching approach facilitates self-framing. The coaching conversation then explores, through a survey, a range of sustainability behaviours. This process allows the participant to consider the mismatch between their normative beliefs and their current behaviours. The dissonance (discomfort) that arises is self-directed and may involve recognition of the gap between behaviour and social expectations, or the personal benefits that may emerge from a behaviour change. The role of the coach is to facilitate this reassessment (barriers and benefits) of the current and alternative behaviour and then to articulate a firm intention or practical plan to undertake the change. The conversation becomes a 'change moment'.

The Coaching for Change framework is an expression of the intervention tools and techniques that influence behaviour change within a model of planned behaviour. The models are consistent, where coaching for change is the trigger of change. The key components of the coaching conversation, set out in Table 10.1, were developed during the second Living Smart programme and further refined through application to the H_2ome

Table 10.1 Key components of the coaching conversation

Coaching technique	Coaches, conversation	Example of participant response
Connect (to heart)/self-framing	How do you feel about conserving water?	It's important to me to do the right thing and to teach my kids
Scope problems (hands/heart)/social normative dissonance	Which of the following water-saving ideas do you currently do? …	Oh, I don't actually have a water-efficient showerhead
Problem (head/heart) personal dissonance	When did you last think you were wasting water?	I always feel bad when I see the kitchen tap dripping
Self-efficacy (Head and Hands)	Have you thought of a way of solving that?	I know my cousin is pretty handy, I just need to make it a priority to ask him to help next time he comes around
Normalize (heart and hands)	Many people have found that (potential solution)	Ok, that sounds like something I could have a go at
Social contract (heart) Accountability (head)	Can I put you down to do that (specify)? Next time we call we chat about how that went and see if your meter readings have come down.	I suppose I'd better not let it slip this time
Experience (hands)	It sounds as if you would get a lot out of doing a water audit of your home.	That would help me to stay on top of things
Collective (social norms) response (heart)	Together participants have already saved …	Wow – I'm part of something that makes a difference

Smart water conservation programme. The language of Head, Heart and Hands is used to define the type of experience that the participant will draw upon to consider the issue raised by the coach. The types of dissonance that emerge are unique to the individual and quite diverse.

On the advice of the lead evaluator, the Department of Transport is planning to run a randomized control trial of Living Smart. This field experiment is designed to test the influence of the information, feedback, coaching and home assessment components of the Living Smart Households programme in a way that is independent of self-selection by different types of participant.

METHODOLOGY

The methodologies tested in the Living Smart Households project draw upon several innovative approaches:

- The broad sustainability scope, goal setting and feedback approach is drawn from the Living Smart™ community course (see http://www.livingsmart.org.au).
- The phased approach to contacting, servicing and motivating households is drawn from the Individualised Marketing™ approach and from health promotion best practice.
- The communication tools are drawn from community-based social marketing.
- The coaching techniques are drawn from a community development approach.

In each component of programme development the Department of Transport engaged the relevant contractor and brought its own expertise and experience to inform the final design. Figure 10.4 was developed by Ben Kent at the Department of Transport to support (successful) applications for awards. It describes the mechanics of the first Living Smart Households project delivery methodology.

The first demonstration project tested a design where sustainability topics were offered to participants in a sequential manner (Topic Design) against offering participants a progressive mix of all topics (Levels Design). Leading personalized marketing contractors were engaged to deliver an information service and motivation-based approach to the first demonstration project. The evaluation revealed no consistent difference in the behaviour change outcomes between the Topic or Levels groups.

The second demonstration project took a blended approach, offering pairs of topics over time and deploying a coaching approach through innovative contractors Sinclair Knight Merz. Figure 10.5, developed by Stephanie Jennings and Colin Ashton-Graham, presents a summary of the deployment of normative techniques into the coaching process.

Both projects involved engagement with households on around six occasions over a duration of a little less than 12 months. The broad

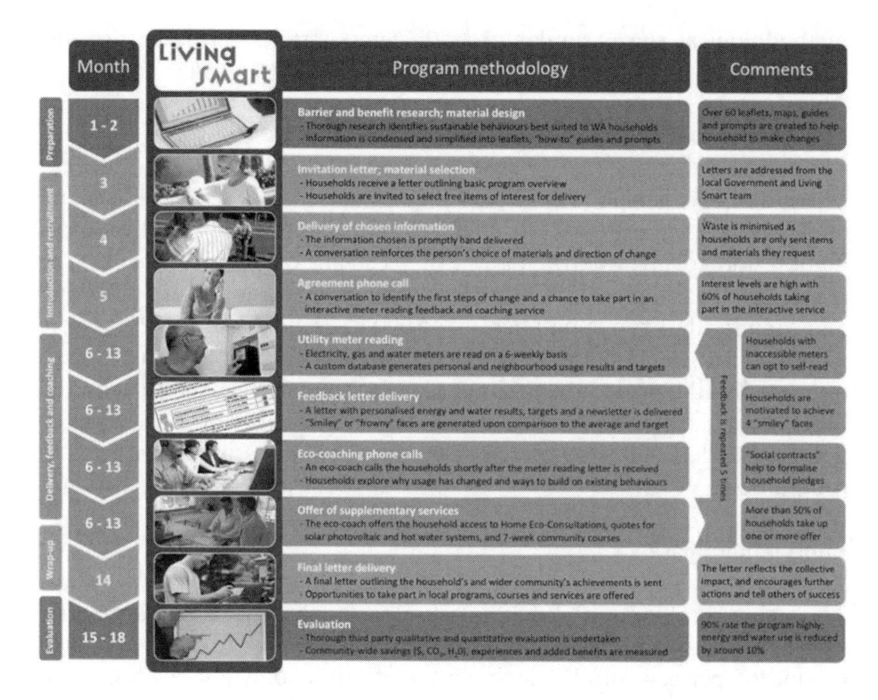

Figure 10.4 Living Smart methodology

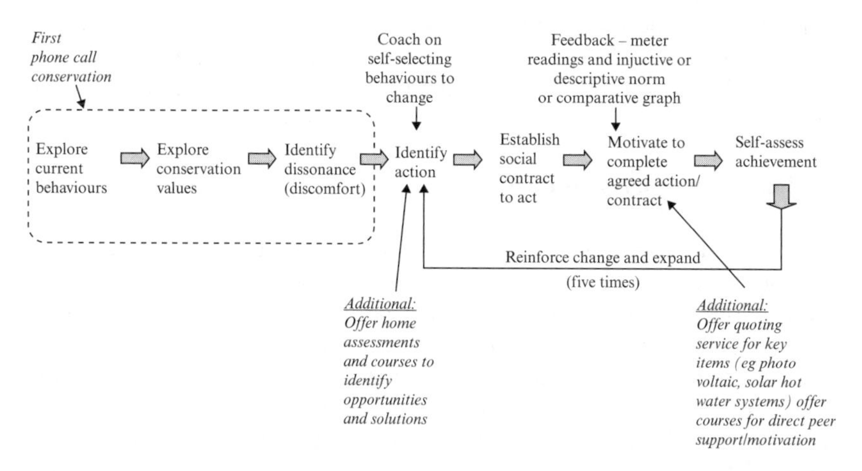

Figure 10.5 Behaviour change model for Living Smart Households

methodological steps, similar in both demonstration projects, are discussed below.

Engagement and Recruitment Phase

The initial engagement with households involved a series of letters and telephone calls to establish an interest in the programme, to offer information and to invite participation in receiving ongoing feedback. The first project was delivered in the Perth metropolitan centres of Joondalup and Mandurah. From a database of 15000 households, some 12000 were identified for telephone contact resulting in 9255 households (80 per cent) interested in the programme. Similar levels of engagement were achieved in the second project, in Perth's eastern region, with 7770 households from a database of 10000 with telephone connection expressing interest in the programme.

Information Phase

Through a combination of mailout/ mailback order forms and outbound telephone marketing calls, orders for information were achieved from around 65 per cent of all telephone target households. On average each participating household, in the first project, ordered around 20 items of information. The second project tailored the information offer more keenly to the households' intended changes and provided an average of ten items per household. In each case the information was individually tailored to the needs of the household. In the second project a coaching conversation was also conducted where many households identified the first changes that they would like to make to their water and energy behaviours.

Feedback (and Coaching) Phase

Some 5943 households agreed to take part in the feedback phase of the first programme (representing almost 50 per cent of the telephone-listed group). Similarly, 4631 households (46 per cent of the telephone group) agreed to take part in the coaching phase of the second programme.

Each of the two programmes provided five rounds of feedback calls based upon a personalized meter reading that reported the level of energy (gas and electricity combined) and water use compared to other households in the neighbourhood. This letter formed the basis for a conversation about intended conservation behaviours. In the first programme the conversation focused on a target level of reduction, while the second programme focused on the identification of specific actions that would lead to a reduction. In each programme the role of subsequent calls was to provide feedback on the degree to which households had reduced their use and/or achieved their actions. In both programmes further information was offered during the series of feedback calls. For some households this was information on the next 'topics' of travel and waste, for others it was additional information on all topics.

Throughout the programme participants were offered access to support services such as home sustainability consultations, access to Living Smart small group courses and (for the second programme only) a quotation service for solar photovoltaic or solar hot water systems. In the first programme almost 28 per cent of the telephone-listed households took up the home consultation compared to almost 21 per cent in the second programme. One difference here was that the first programme included a short visit to install free energy-efficient globes; this service was not available for the second programme. The effect of the differences between the two programmes is explored in the monitoring findings that follow.

MONITORING

A sophisticated monitoring programme was put in place including: internal project Key Performance Indicators (KPIs), travel diaries, an independent quality survey and meter readings (gas, electric and water) over a three-year period.

One of the most interesting innovations in the second Living Smart Households project was the sophisticated customer database enabling the tracking of reported sustainability behaviours. During the recruitment and coaching phases the behaviours of more than 6000 households were coded as 'Pre-existing', 'Living Smart Agreed', 'Living Smart Achieved' or 'Not Applicable/Possible'. This information (Figure 10.6) was utilized to provide households with normative feedback on the uptake of actions by others and to calculate the likely energy and water savings across the community at the end of each coaching round.

The coaching records reported 7782 and 4061 energy and water-saving actions, respectively. This equates to an estimated energy saving of 14000 kWh (equivalent to 3000 1 kW peak photovoltaic systems or 2.9 kWh per household per day) and 197 kL (40 L per household per day) per day of water saving. An initial analysis of electricity data shows savings of 1.7 kWh per household per day (around 8 per cent saving) at the mid-point of the project. Further tracking of project outcomes are subject to budget availability.

The first project only produced marketing (for example, materials

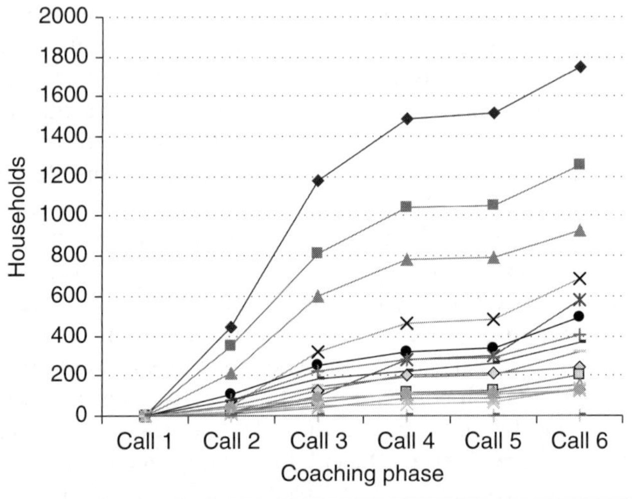

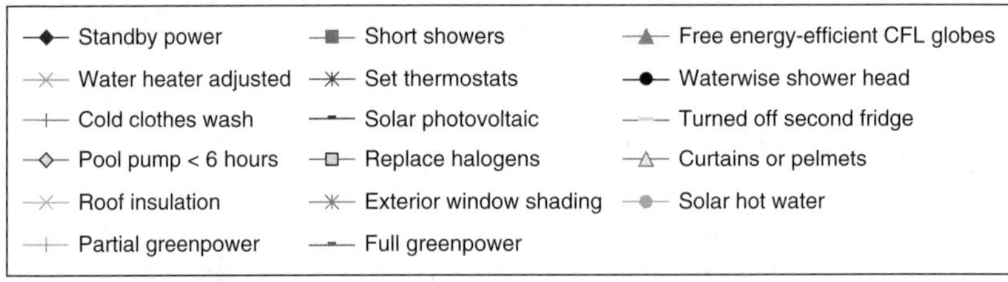

Figure 10.6 Energy actions

ordered) indicators and no behavioural indicators. However, a post-programme quality survey was conducted with a sample of 220 households from the first project. This research provided useful measures of self-reported behaviours. Figure 10.7 shows the proportion of Living Smart Households' participants who reported making changes in each of the main Living Smart topic areas.

Some of the findings of the face-to-face surveys with participants are as follows:

- 86 per cent found the home consultations very or quite useful.
- 85 per cent found the meter reading letters, newsletters and feedback phone calls very or quite useful.
- 80 per cent agreed that the information provided encouraged them to change their electricity and/or gas usage.
- 72 per cent agreed that the information provided encouraged them to change their water usage.

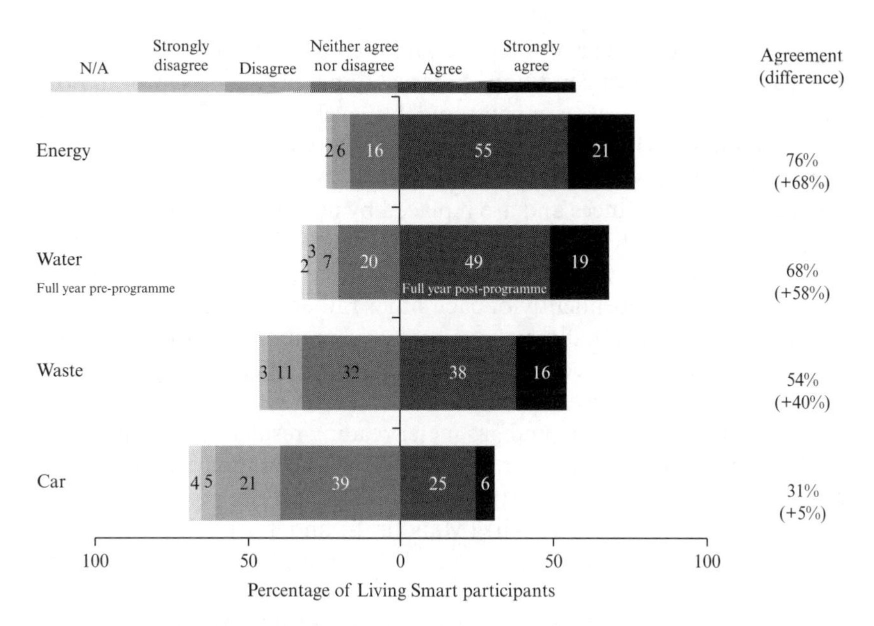

Figure 10.7 Self-reported behaviour changes (first Living Smart Households project)

- 31 per cent agreed that the information provided encouraged them to reduce their car trips.
- 84 per cent of those who attended found the community workshops and courses very or quite useful.

A similar quality survey was conducted with a sample of participants from the second project. Similarly, 89 per cent of responses across nine quality measures rated the programme highly. Likewise, 87 per cent of participants believed that the programme had helped them to reduce their energy use, 86 per cent to reduce water use, 57 per cent to reduce waste and 35 per cent to reduce car use.

Independent and Meter Data

The first Living Smart Households project has now accumulated full year pre- and post-programme meter reading data for energy (electricity and gas) and water consumption. The dataset includes more than three million individual meter readings stored in a depersonalized file and coded by participation type and control group status. This data were provided to the University of Western Australia for analysis.

Target (participant group) and localized control group data were analysed across all project conditions (intervention types such as information, coaching and home assessments) and by level of use of each resource (high/medium/low). Regression to the mean (where high or low consuming households tend towards the average over time due to changing household circumstances and are replaced by new high or low users experiencing the opposite demographic or structural changes) was observed and controlled for. The evaluation approached gas, electric and water meter readings independently of each other rather than through a regression model approach that may have produced multicollinearity errors due to participation being a common determinant of reductions in gas, electric and water consumption.

Evaluation of electricity and gas meter reading results by the University of Western Australia (full year pre- and post-Living Smart Households) revealed that programme participants reduced energy consumption by between 1 kilowatt hour (kWh) (Mandurah) and 1.9 kWh (Joondalup) per household per day (between 3.4 and 5.4 per cent of household energy use).

This is evidenced through electricity savings of 0.5 kWh and 1.3 kWh per household per day in Mandurah and Joondalup, respectively (about 3 to 6 per cent of household electricity use) and gas savings of 0.5 kWh (Mandurah) and 0.7 kWh (Joondalup).

Figure 10.8 shows that electricity use declined for the participants in Joondalup between 2007–08 (full year pre-programme) and 2009–10 (full year post-programme) and increased over the same period for the nearby

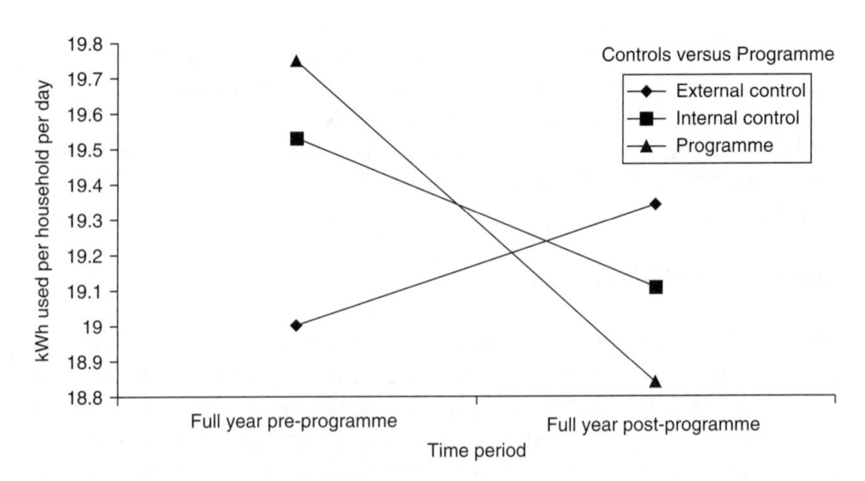

Figure 10.8 Change in electricity use in Joondalup

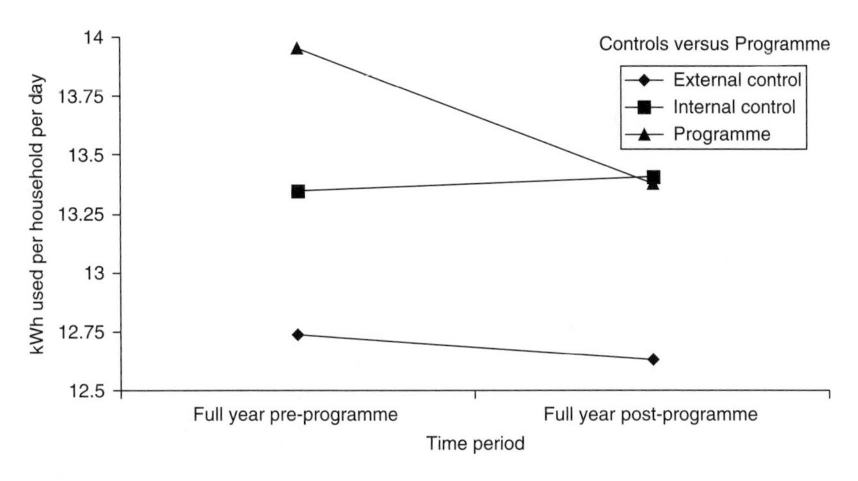

Figure 10.9 Change in gas use in Mandurah

(external) control group. Similar results were found for electricity use in Mandurah, for gas use in Joondalup and gas use in Mandurah Figure 10.9.

Similar analysis of water meter data found that there was a variance in water savings across the seasons. Participating households from the City of Mandurah saved 53 litres of water per day (6 per cent) in summer (July to February), while those from the City of Joondalup saved just 23 litres per day (3 per cent) in winter (April to November). The savings in water were calculated for comparable seasons (part year only) prior to the programme (2007–08), during the programme (2008–09) and post-programme (2009–10). Figures 10.10 and 10.11 present the data for Mandurah and Joondalup water use, respectively (95 per cent confidence intervals are shown for each mean measure). Further analysis of water data, to measure the savings a full year after the programme, was not possible due to a change in the meter reading dates in some areas introducing an uncontrolled seasonal influence.

Before and after surveys of a random sample of participants were conducted by the contractor (Socialdata) in order to determine the effect of the programme on travel patterns. In Mandurah, twin random samples of 553 (before) and 490 (after) households provided travel data for all persons in each household for one day (representative of all days). In Joondalup twin random samples of 579 (before) and 425 (after) households were collected. In both analyses the whole group of target households were analysed, including non-participants in the Living Smart service.

Across the whole target group the number of trips by car dropped by 5 per cent, with walking increasing by 18 per cent, cycling by 4 per cent

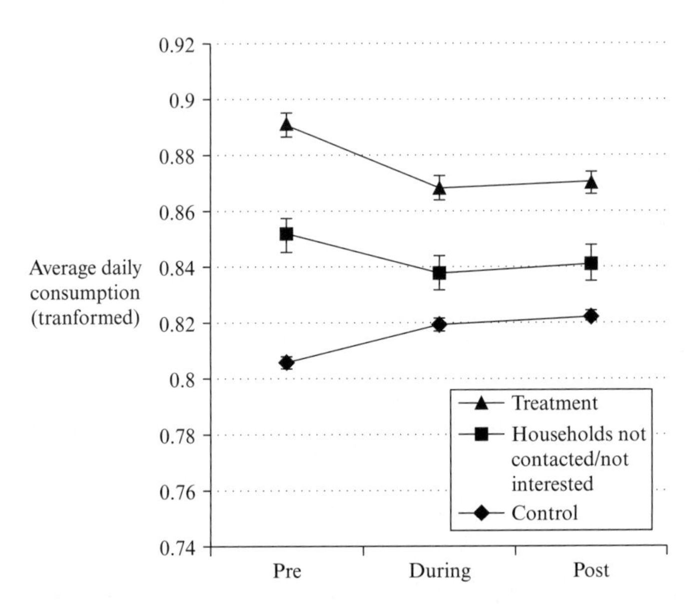

Figure 10.10 Daily water use – Mandurah

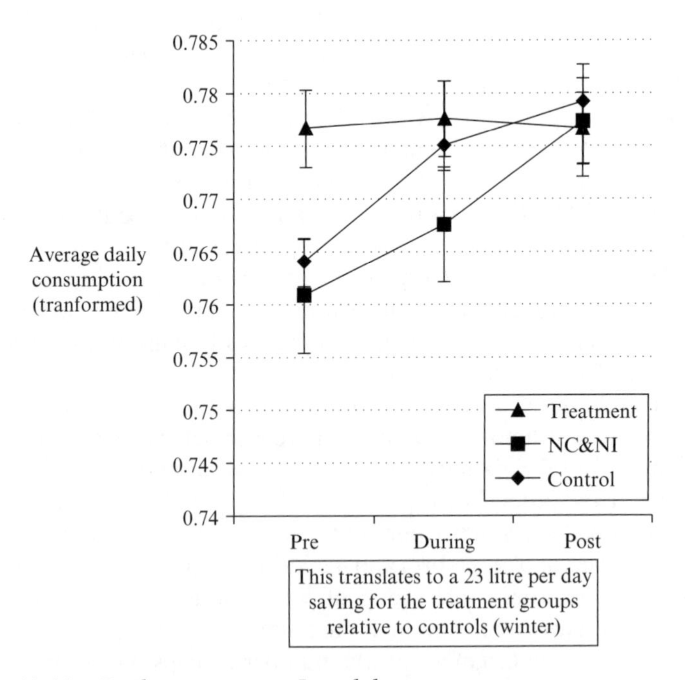

Figure 10.11 Daily water use – Joondalup

Table 10.2 Change in trips per person per year in Mandurah after rail and with Living Smart

Modes	2007 before rail	2008 after rail	Relative change before/ after rail	2010 after Living Smart	Relative change with rail/with rail and Living Smart
	Trips per person per year	Trips per person per year	Change %	Trips per person per year	Change %
Walk	76	82	8.4	97	18.3
Cycle	23	24	4.9	25	4.2
Motorcycle	6	8	34.0	4	–50.0
Car driver	608	601	–1.1	570	–5.2
Car passenger	261	245	–6.0	251	2.4
All public transport	22	35	59.9	46	31.4
Total	995	995		993	

Table 10.3 Change in trips (Joondalup Living Smart areas)

Per person per year	Before	After	Relative change %
Walking	88	102	+16
Bicycle	17	22	+29
Motorcycle	3	3	N/A
Car as Driver	676	647	–4
Car as Passenger	276	262	–5
Public Transport	44	47	+7
Total	1104	1083	

and public transport use by 31 per cent. These increases in sustainable transport choices came on top of the 60 per cent increase in public transport trips and 1 per cent drop in car trips following the opening of a new Mandurah line rail service (Table 10.2).

Similar results were measured for Joondalup where there were no major changes to the transport system (Table 10.3). In both cases the overall levels of activity (that is, tasks such as work, shopping and so on) were unchanged, although trip chaining (combining errands) resulted in a small reduction in the total number of trips made. The major change observed was in switching from the car to alternative modes of travel.

Preliminary data analysis suggests that the second Living Smart Households project has performed better on energy reduction and similarly in water and car travel reductions. Internal analysis conducted by the energy network provider, for instance, demonstrates an electricity reduction of 8 per cent (1.3 kWh per household per day) amongst participants in the second Living Smart Households project (Western Power, 2011). Pre- and post-programme gas data was not available for this project. Analysis of water consumption data for 12 months prior to the programme compared to 12 months during/post-programme demonstrates water savings of between 28 litres per household per day (3.6 per cent compared to the internal control group of neighbouring households) and 68 litres per household per day (8.7 per cent compared to the baseline).

A pre- and post-travel diary survey, conducted by independent analyst Synovate, shows a 4.3 per cent reduction in car driver trips amongst participating households as compared to a control group. The travel diary survey fieldwork was interrupted by a severe hailstorm in the before survey period and by inclement weather during the after survey. Not all indicators, by mode and distance, were consistent or logical and hence the travel result from the second Living Smart project is unclear.

EVALUATION

The results have been analysed to estimate a cost per tonne of carbon abatement and to compare the behavioural approach to the cost of infrastructure/supply-side alternatives. The approach is shown (below) to be cost-effective for the business case of utilities and for emerging carbon markets.

Across both Living Smart Households projects pre- and post-data reveal a 4–8 per cent reduction in energy and water consumption for the target group compared to the control and 4–5 per cent reduction in car use. A list of self-reported behaviour change was collected through the coaching and quality survey process revealing a large uptake of actions on standby power, fixing leaking taps and so on. The behaviour changes approximate to 1 tonne of carbon dioxide-equivalent (CO_2-e) per participating household per annum. Participation rates of approximately 50 per cent of the local population makes the collective impact significant (Government of Western Australia, 2012b).

Travel diary survey responses indicated a reduction in car travel of 3 kilometres (km) per household per day. This equates to an annual saving of around 1000 km and 300 kg of CO_2-e per household per annum. Participating households experience a reduction of 90 litres of petrol used per household per year, and a saving of around $AUS 120, based on fleet average city fuel cycle of 0.3 kg CO_2-e /per km, 9l per 100 km consumption and $1.30 per litre (Australian Government Bureau of Infrastructure, 2008).

Based on self-reported actions and meter reading validations, an energy saving (gas and electricity combined) of around 2.0 kWh per participating household per day is estimated (about 6 per cent of household energy use). This translates to an annual saving of around 700 kWh per household, equivalent to a reduction of around $AUS 130 on annual energy bills and a saving of 560 kg of CO_2-e per household per annum (based on 500 kWh electricity at 0.9 kg CO_2-e per kWh and $AUS 0.22 per kWh and 200 kWh gas at 0.3 kg CO_2-e per kWh and $AUS 0.12 per kWh). With some 14000 participating households, or more than 60 per cent of the households invited to join the projects, more than 12000 tonnes of carbon abatement was realized in the first year of the programme.

Water savings of around 50 litres per household per day (7 per cent) equate to water use charge savings of around $AUS 20 and a CO_2-e saving of around 18 kg per household per annum (based on 1 kg CO_2-e per kL and $AUS 1.19 per kL). While these savings (amounting to $AUS 270 and 880 kg of CO_2-e per annum) are modest on a per household basis, the large reach of the programme results in both significant and cost-effective outcomes.

On a per household basis the project costs (all contractors, materials and government project management) around $AUS 285 to deliver. This one off cost results in a first year benefit to households (reduced bills) of $AUS 270 and an abatement cost of $AUS 323 per tonne of CO_2-e (based on 2011 energy and fuel costs). The carbon abatement cost will be reduced as annual reductions accumulate and if the value of infrastructure and supply costs avoided are deducted from the project cost.

A simple (arbitrary) allocation of project costs equally across transport, water, energy and carbon abatement results in project costs that are competitive as an alternative to infrastructure alternatives (Table 10.4).

In addition to the direct demand-side benefits the Living Smart Households approach provides secondary benefits such as increased physical activity (active transport modes) and associated health benefits, improved community connection and increased engagement in policy responses to climate change.

While a full socio-economic evaluation has yet to be conducted, the cost sharing achieved by the Living Smart model suggests that it will outperform the Travel Smart intervention for which a socio-economic benefit:cost ratio of around 30:1 has been demonstrated (Ker, 2002). From a policy perspective the very large (30:1) benefit:cost ratio offers sufficient

Table 10.4 Comparative costs of demand reduction through Living Smart with supply-side alternatives

Policy benefit	Cost per household	First year cost of outcome	Five year cost of outcome (assumed 20% decay)	Comparative supply-side cost
Water	\$71	18.3 kL saved at \$3.88 per kL	\$0.97 per kL	\$2.00 per kL desalination supply (marginal)
Energy	\$71	700 kWh saved at \$0.10 per kWh	\$0.03 per kWh	\$0.18 per kWh average cost of supply
Transport	\$71	1095 km avoided at \$0.06	\$0.01 per km	no benefit cost in place
CO_2	\$71	0.88 tonnes avoided at \$80.68	\$12.90 per tonne CO_2	\$23 per tonne CO_2 Australian carbon price 2012–15

margin to accommodate the greater risks of failure associated with behavioural interventions when compared to infrastructure provision.

LEARNINGS

As an applied programme to test behaviour change methodologies, the work of the Western Australian Government has been pioneering. The Living Smart Households programme has demonstrated an innovative set of approaches and a strong application of learning by doing. The second Living Smart Households programme in particular has demonstrated success in delivering, through application of personalized social marketing techniques, cost-effective and large-scale behaviour change. The key learnings are that:

- An engagement that facilitates 'self-framing' is powerful for recruiting a high proportion of the target population and engaging them in the process of change.
- The 'dissonance dose' (level of discomfort) that is experienced by households is critical to the execution of change.
- The combination of 'self-framing' and 'dissonance dose' work together to challenge and support households.

- Feedback based on comparative measures and social norms are a powerful motivator of change.
- The delivery methodology of 'coaching' provides a mechanism for households to receive feedback (on meter readings) and then to define a 'change moment' sufficient to put positive strategies in place.
- Coaching techniques that include a 'social contract' result in high completion rates for nominated changes.
- High quality, and fully personalized, service delivery sets up a 'reciprocity' through which households feel motivated to 'do the right thing'.
- The combination of information, feedback, coaching and in-home services is effective.
- The combined (water, energy, travel and waste) approach of Living Smart Households is cost-effective.

FUTURE CHALLENGES

As a 'natural experiment' the Living Smart Households projects have not fully controlled for the effect of each of the project components nor for the effect of self-selection into the programme or services. The Department of Transport has commenced a project to randomly assign households to combinations of information, feedback, coaching and home services. The outcome of this 'controlled experiment' will be clarity on the relative effect, and value for money of each component of the service.

As an innovative programme there is no benchmark for success. Further testing of messages, feedback styles, community engagement and target demographics will be required to extend the learning by doing approach.

The structure of government and the energy and water utilities presents the greatest challenge to harnessing the efficiencies of a sustainability service for households. A combination of 'silo' policy thinking and competition between private sector suppliers acts to impede the genuine 'joined up thinking' that is evidenced in the Living Smart Households programme. Living Smart Households could provide cost-effective abatement for carbon price liable entities such as waste managers (land fill emissions), energy producers and major energy consumers (such as water providers who pay the increased cost of carbon-priced energy). As a carbon price adds significant cost to supply-side responses and the carbon price becomes a common currency across portfolios, Living Smart Households may turn out to be a breakthough in behaviour change for sustainability.

REFERENCES

Ajzen, I. (2006), 'Constructing a theory of planned behaviour questionnaire', available at people.umass.edu/aizen/pdf/tpb.measurement.pdf (accessed 18 December 2012).

Ampt, E. and C. Ashton-Graham (2008), 'One at a time or all at once? Experience in the uptake of environmental behaviour change measures', *Proceedings of the International Solar Cities Congress* Adelaide, available on a CD-ROM and at http://Catalogue.nla.gouam/Record/4656589:loohfor=anthor:%2International%20Solar%20Cities%20Congress%202008%20:%20Adelaider%20S.%20Aust.)%22Shoffset=1&max=1 (accessed June 2013)

Ashton-Graham, C. (2008), 'TravelSmart and Living Smart case study', available at www.garnautreview.org.au/ca25734e0016a131/WebObj/Casestudy-TravelSmartandLivingSmart-WesternAustralia/$File/Case%20study%20-%20TravelSmart%20and%20LivingSmart%20-%20Western%20Australia.pdf (accessed 18 December 2012).

Ashton-Graham, C. and G. John (2006), 'Travel Smart household program: frequently asked questions in travel demand management and dialogue marketing', *Government of Western Australia, Department of Transport*, available at www.transport.wa.gov.au/AT_TS_P_faqs.pdf (accessed 18 December 2012).

Ashton-Graham, C. and H. Putland (2007), 'Couch potato or gym junkie', Be Active Conference, Adelaide, available at www.transport.wa.gov.au/media Files/AT_TS_P_couchpotato.pdf (accessed 18 December 2012).

Australian Government Bureau of Infrastructure, Transport and Regional Economics (2008), *Australian Transport Statistics June 2008*, available at www.bitre.gov.au/publications/2008/files/stats_019.pdf (accessed 17 December 2012).

Byrne, J. (2006), *The Green Gardener*, New York: Penguin Books.

Cairns, S. (2004), 'Smarter choices – changing the way we travel', available at www.dft.gov.uk/publications/smarter-choices-changing-the-way-we-travel-main-document/# (accessed 18 December 2012).

Costa, D. and M. Kahn (2010), 'Energy conservation "nudges" and environmentalist ideology', available at www.nber.org/papers/w15939 (accessed 17 December 2012).

Garnaut, R. (2008), *The Garnaut Climate Change Review*, Cambridge: Cambridge University Press.

Government of Western Australia, Department of Transport (1995), *Metropolitan Transport Strategy*, available at www.transport.wa.gov.au/AT_TS_P_Perth MetroTransportStudy.pdf (accessed 18 December 2012).

Government of Western Australia (2007), *Making Decisions for the Future: Climate Change*, available at portal.environment.wa.gov.au/. . ./2007006CLI MATECHANGE.PDF (accessed 18 December 2012).

Government of Western Australia, Department of Transport (2012a), *Living Smart: Acting on Climate Change*, available at www.transport.wa.gov.au/AT_LS_P_acting_on_climate_change.pdf (accessed 18 December 2012).

Government of Western Australia, Department of Transport (2012b), *Living Smart Households (Sustainability Program) – Monitoring and Evaluation*, available at www.transport.wa.gov.au/mediaFiles/AT_LS_P_Monitor_Eval_Summary_Report.pdf (accessed 18 December 2012).

Ha, T. (2003), *Greeniology*, New York: Allen & Unwin.

Ker, I. (2002), 'Preliminary evaluation of the financial impacts and outcomes of the TravelSmart individualised marketing program – update', available at www.transport.wa.gov.au/AT_TS_P_financereport.pdf (accessed 18 December 2012).
Kollmuss, A. and J. Agyeman (2002), 'Mind the gap: why people act environmentally and what are the barriers to pro-environmental behaviour?', *Environmental Education Research*, **8** (3), 239–60.
McKenzie-Mohr, D. (2010), 'Fostering sustainable behaviour', available at www.cbsm.com/pages/guide/preface/ (accessed 18 December 2012).
Newman, P. (2005), 'Sustainability in the wild west', in K. Hargoves and M.H. Smith (eds), *The Natural Advantage of Nations*, Abingdon: Earthscan Books, pp. 271–82.
Newman, P. (2012), 'The Perth rail transformation', CUSP Discussion Paper, available at www.sustainability.curtin.edu.au (accessed 21 June 2013).
Newman, P. and J. Kenworthy (1999), *Sustainability and Cities*, Washington, DC: Island Press.
Nolan, J., W. Schultz, R. Cialdini, N. Goldstein and V. Griskevicius (2008), 'Normative social influence is underdetected', *Personality and Social Psychology Bulletin*, **34**, 913–23.
RED 3 (2006), *Evaluation of Australian TravelSmart projects in the ACT, SA, Queensland, Victoria and WA from 2001–05*, Carberra: Department of Environment and Heritage, Government of Western Australia.
Robinson, L. (2010), 'How to change the world', available at www.enablingchange.com.au/ (accessed 18 December 2012).
Sheehy, L. (2003), 'Evaluation report of the Living Smart pilot program', available at www.livingsmart.org.au/pdfs/03Aug12LivingSmartEvaluationReport.pdf (accessed 27 January 2012).
Socialdata (2000), *Potentials Analysis Perth*, available at www.transport.wa.gov.au/activetransport/24690.asp (accessed 18 December 2012).
Socialdata (2001), 'Travel Smart program contract variation 4', unpublished manuscript, Department of Transport, Government of Western Australia.
Socialdata (2007), *Climate Change In-depth Perth*, available at www.transport.wa.gov.au/mediaFiles/AT_LS_P_survey.pdf (accessed 18 December 2012).
Socolow, R. and R. Sonderegger (1976), *The Twin Rivers Program on Energy Conservation in Housing – Four Year Summary Report*, Princeton, NJ: Princeton University Press.
Water Corporation (2009), *Annual Report*, available at www.watercorporation.com.au/_files/PublicationsRegister/6/2009AR_FULLVERSION.pdf (accessed 18 December 2012).
Western Australia Sports Federation (2010), *Annual Report 2009/10*, available at www.wasportsfed.asn.au/downloads (website no longer available).
Western Power (2011), *Perth Solar City Annual Report 2011*, available at www.perthsolarcity.com.au/annual-report (accessed 18 December 2012).

11. Energy demand implications of structural changes in India

Subhes C. Bhattacharyya

INTRODUCTION

As the Indian economy has abandoned the traditional 'Hindu' growth rate for a faster economic growth path in recent years, major changes to the traditional Indian way of life are emerging. The rapidly growing Indian population, accelerating rate of urbanization, growing size of the middle class and an increasing level of integration of the country with the rest of the world are contributing to these changes. This transformation will have significant implications for both its long-term future and on the structure of the country's energy system. The purpose of this chapter is to analyse the implications of ongoing macro-level changes on energy demand with a particular emphasis on household demand.

While a number of studies have looked at the influence of changing lifestyles on energy demand in India (de la Rue du Can et al., 2009a, 2009b; Filippini and Pachauri, 2004; Pachauri, 2004; World Bank, 2008), they have either focused on a specific area or sector, for example, the McKinsey Global Institute (MGI) analysis of urbanization (2012) or the World Bank's assessment of residential electricity demand (2008), or used an older data set (de la Rue du Can et al., 2009; who used the National Sample Survey Office data of 2001 or 2004–05; Filippini and Pachauri, 2004; Pachauri, 2004). This chapter relies on more recent data to capture the effects on residential energy demand by disaggregating rural and urban populations and considering their income differences.

The chapter begins with a brief review of factors influencing the changes. This is followed by a brief review of relevant literature and a description of the methodology. The analysis of residential electricity demand is then presented. Finally, some concluding remarks are given. It should be highlighted that the chapter considers national-level changes in the demography, urbanization and income distribution patterns and provides an insight into the changes in residential energy demand. It does not try to capture either the differences at the state/province level or any change in individual behaviour at the micro level.

OVERVIEW OF INDIA'S TRANSFORMATION AND BEHAVIOURAL CHANGE

India, the second most populous country in the world following China, has seen her population grow fourfold between 1901 and 2011, moving from 238.4 million in 1901 to 1021 million by March 2011 (Government of India, 2012). Although India added 181 million people between 2001 and 2011, the average growth rate is showing a declining trend compared to previous decades. It will take a considerable amount of time before India can stabilize her population to reach the fourth stage of the demographic transition. According to the United Nations population forecasts, India's population will grow to 1.46 billion by 2025, and to 1.69 billion by 2050, thereby increasing her share to more than 18 per cent of the global population (United Nations, 2011). India is projected to overtake China in population count by 2021 and become the most populous country in the world. This assumes importance as India occupies only 2.4 per cent of the global surface area but supports more than 17 per cent of the global population (Government of India, 2012).

Demographic Transition

Alongside the demographic transition, the structure of India's population is also changing (Figure 11.1). The size of the working-age population grew significantly between 1950 and 2010 and the same trend is projected to continue over the next three decades. This indicates that the ratio of working-age population to that of non-working age population will improve and peak around 2040 (Bloom, 2011). The change in population structure has the potential to a demographic dividend for the country. Bloom (2011) suggests that if India uses her working-age population productively, there is potential for benefiting from an extra percentage point of growth in gross domestic product (GDP) per capita.

It is worth highlighting that about 1 per cent of the Indian population lives overseas (World Bank, 2010) with the majority of these emigrants being skilled workers. Since 1990, the remittances from overseas Indians have increased exponentially, and in 2010, India received $55 billion, representing the highest volume of capital flow into the country in that year. India also emerged as the highest recipient of overseas remittances from its overseas population. The liberalization of the economy in the

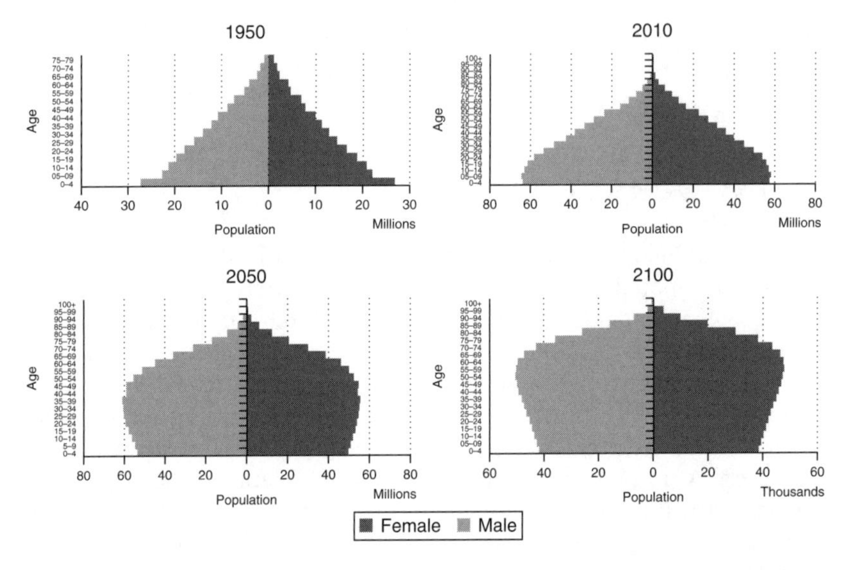

Source: World Population Prospects: The 2012 Revision, http://esa.un.org/unpd/wpp/index.htm.

Figure 11.1 Population by age group and gender (absolute numbers)

1990s and the increased flow of professionals going abroad since 1990 (particularly information technology professionals) have contributed to the growing remittance volume (Singh and Hari, 2011). Such capital flow influences consumption behaviour and helps improve living conditions of the recipients.

Accompanying the population growth has been a significant increase in urbanization. According to the United Nations (2010), the Indian urban population grew to 364 million in 2010 from 63 million in 1950, meaning that about 30 per cent of the Indian population lived in urban areas in 2010. Once again, the growth in the urban population is much higher in recent decades compared to the 1960s and 1970s. Since 1990, India has added 144 million urban dwellers, one result being that the urban population is contributing to an ever-increasing share of the economic growth of the country. MGI (2012) indicates that in 1995, the urban rural share in the GDP was almost equal; by 2008, the urban share had increased to 58 per cent of India's GDP.

The United Nations estimates that the share of the urban population will reach 40 per cent by 2030 and 54 per cent by 2050. This implies that another 226 million people will be added to the country's cities within the next two decades, and more than 510 million by 2050 (Figure 11.2). This

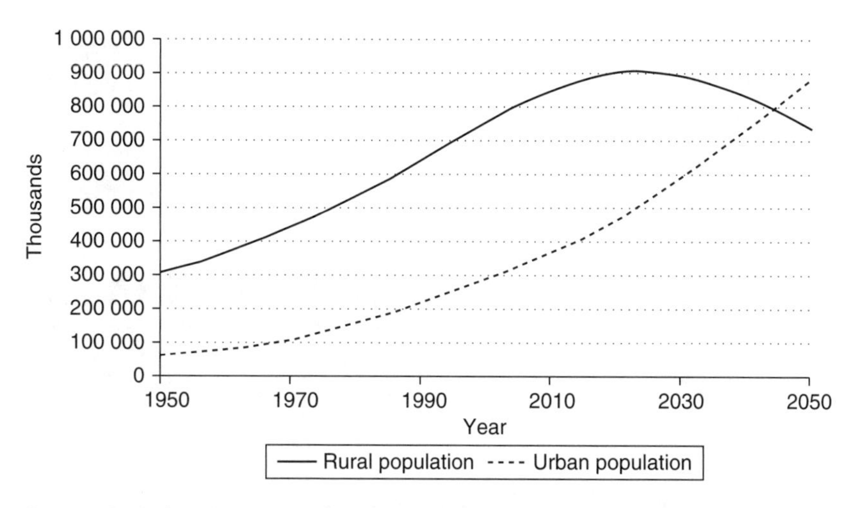

Source: United Nations Population Division, http://esa.org/unpd/wap/unnp/index_panel3 html.

Figure 11.2 Indian urbanization prospects

rate of urbanization is unprecedented in the world, except of course for China, and will have serious implications for the country and the world. MGI (2010) indicates that 70 per cent of India's GDP in 2030 is likely to come from its urban areas and has the potential to drive a fourfold increase in the per capita income by 2030. Simultaneously, the size of the middle-class population (earning between 200 000 and 1 million rupees per year) will grow fourfold to 147 million from the present 2012 level of 32 million.

Rural–urban Consumption Differences

The increasing rate of urbanization as well as the improving skills of the working-age population will directly affect the consumption decisions of India's urban population. According to the most recent National Sample Survey Office (NSSO) on household consumption (2012), an urban consumer was likely to have spent almost twice that of his rural counterpart in 2009–10; the spending pattern was also reported to be significantly different between the two (Figure 11.3). In monetary terms (rupees per capita per month), an urban consumer spends 95 per cent more than his rural counterpart. In terms of food items, an urban consumer spends about 46 per cent more than his rural colleague while the spending disparity in non-food items is even more pronounced (more than 260 per cent higher). This

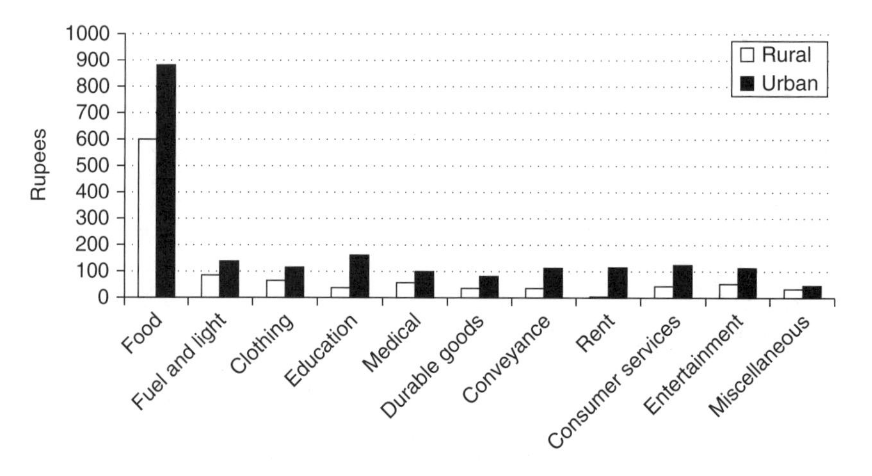

Note: The expenditure is presented in rupees per person per month.

Source: NSSO (2012). The NSSO carries out surveys at the national and state levels to collect primary data on living conditions. The 66th round focused on household consumption and employment issues.

Figure 11.3 Rural–urban expenditure difference in India

is due to a high level of spending on rent (for accommodation), education, transport and durable goods. As shown in Figure 11.3, since the spending pattern is likely to reflect the differences in the income level in rural and urban areas, the sample data point to significant rural–urban disparity existing in the country.

The difference in the share of different items in the expenditure basket reveals that a rural consumer spends 57 per cent of his budget on food while the urban consumer spends about 44 per cent of his budget on food (Figure 11.4). Clearly, the food-related expenses represent a major element in both categories of consumers. Similarly, education and energy-related costs emerge as the next most important items for urban consumers. Rent, clothing, transport, other consumer services, entertainment and medical expenses get very similar priorities in the consumption spending decision. Durable goods come at the end of the spending list. For the rural consumer, on the other hand, energy spending represents the second most important item, while clothing gets the third highest priority. Medical expenses, rent and entertainment have similar shares in the rural budget, while education, consumer services and durable goods receive less priority in terms of spending. Clearly, the higher priority given to education for urban dwellers is also a distinctive feature. The recognition of better

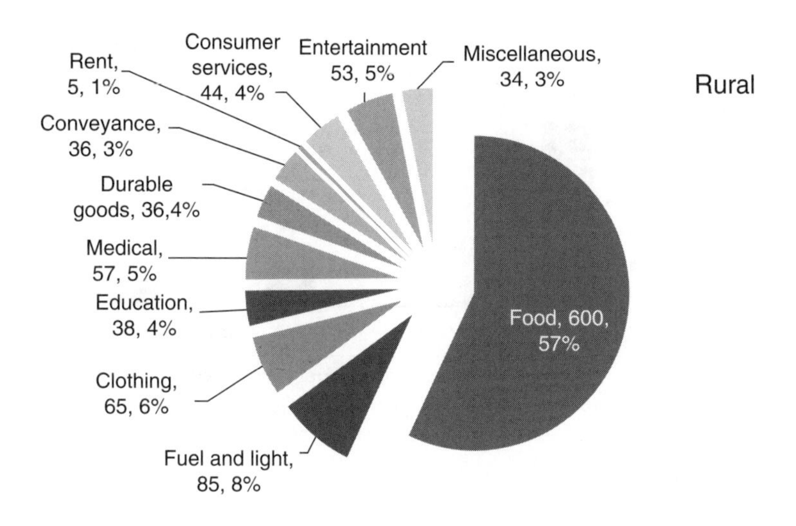

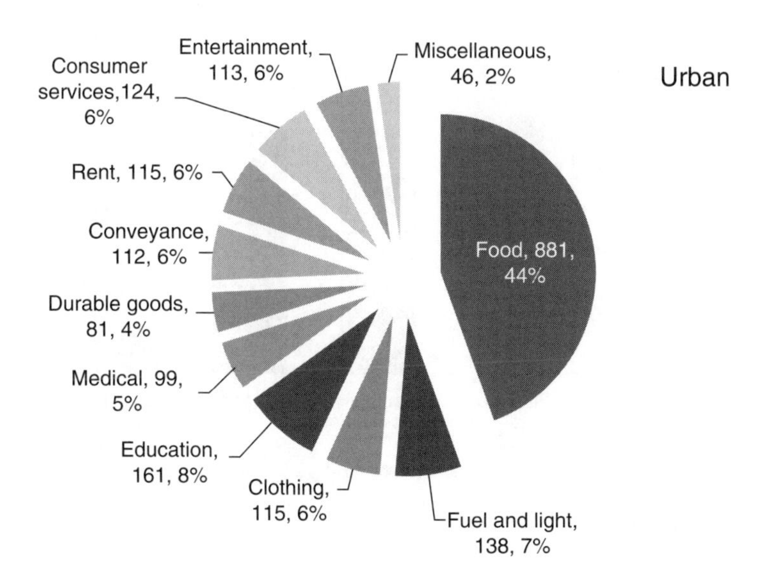

Note: Expenditure figures are given in rupees per person per month for 2009–10 (first number). Percentages shown are the percentage share of expenditure in total.

Source: NSSO (2012).

Figure 11.4 Differences in expenditure patterns between rural and urban India

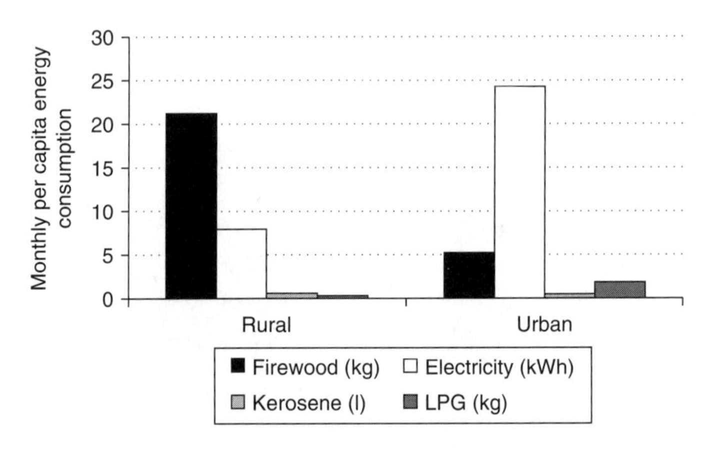

Source: NSSO (2012).

Figure 11.5 Disparity in energy use between rural and urban India

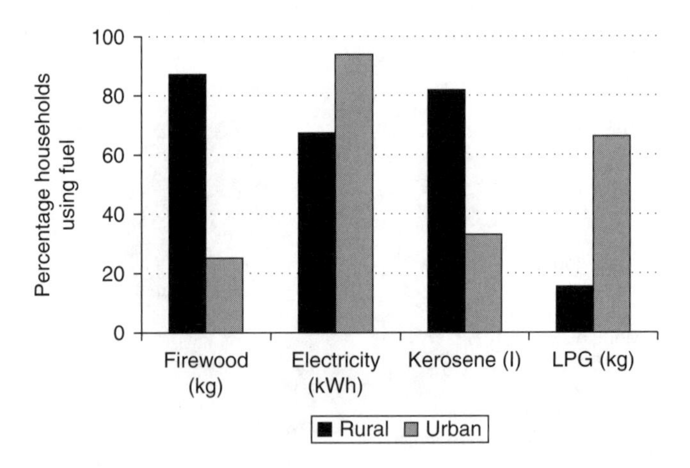

Source: NSSO (2012).

Figure 11.6 Share of households using different energies

education for a better future, and higher peer pressure and competition in education can explain the high level of expenditure in this regard.

A closer look at some of the specific items such as energy, durable goods and transport-related expenditure reveals a number of interesting features (Figures 11.5 and 11.6). While rural consumers still rely heavily on firewood and solid wastes for their energy needs, urban counterparts rely more on modern forms of energy. The fuel choice reflects the level of access to modern energies as well as consumers' ability to pay for a fuel and related government interventions. Rural households still rely heavily on firewood (87 per cent of rural households use firewood) for cooking, with a per capita consumption of about 21 kg of firewood per month while the urban households have moved towards liquefied petroleum gas (LPG) (66 per cent urban households use LPG). The urban consumers are using about 5 kg of firewood and 1.8 kg of LPG per person per month. In terms of electricity use, there is a wide disparity as well: 94 per cent of urban households have reported electricity use while 69 per cent of rural households on average consume electricity, indicating access constraints in rural areas. While an urban consumer uses about 24 kilowatt hours (kWh) per month, a rural consumer consumes only about 8 kWh per month. Limited affordability of rural consumers and limited ownership of electric appliances as well as supply constraints explain the difference in rural–urban electricity consumption behaviour.

Moreover, for transport purposes, NSSO (2012) reported a significant change in petrol use in India. The number of households reporting use of petrol for transport purposes has increased by 81 per cent in rural areas between 2004–05 and 2009–10, while just a 25 per cent increase is reported in urban areas (Figure 11.7), although in absolute terms the

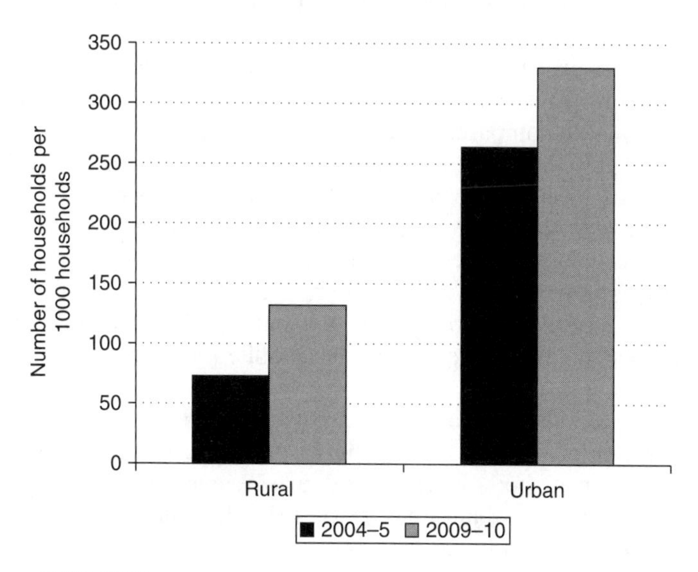

Source: NSSO (2012).

Figure 11.7 Use of petrol for transport purposes

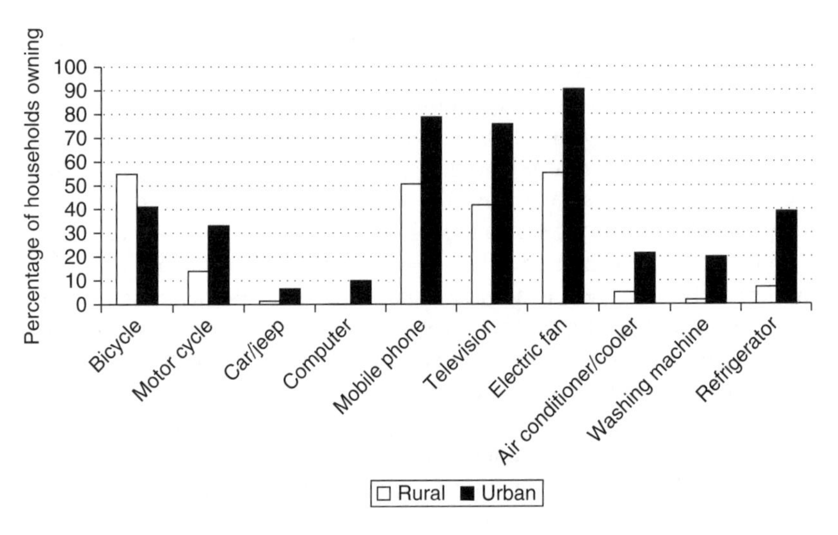

Source: NSSO (2012).

Figure 11.8 Ownership pattern of durable goods

picture is very different, that is, only 13 per cent of rural households are using petrol compared to 33 per cent of urban households using the fuel.

The ownership pattern of durable goods in urban and rural areas is also revealing (Figure 11.8). More than one half of rural households possess a bicycle compared to about 41 per cent of urban households. On the other hand, urban households own more motorized vehicles compared to rural counterparts. In fact, almost 33 per cent of urban households now own motorcycles (two-wheelers) whereas only 6.5 per cent possess a four-wheeled vehicle. Similarly, televisions, electric fans and mobile phones have become common goods for most of the households both in rural and urban areas. A high level of ownership of mobile phones even in rural areas reflects the mobile phone revolution in the country.

However, a closer look at the vehicle ownership pattern by expenditure decile class (Figure 11.9) reveals that bicycles are owned by relatively poorer consumers both in rural and urban areas but the ownership declines very fast in urban areas beyond the third decile class. In turn, motorized two-wheeler ownership increases rapidly as consumers' income grow. For example, 9 per cent of households in the second decile are likely to own a two-wheeler but this rises to 55 per cent in the ninth decile class. The richest income class, on the other hand, is likely to prefer cars/four-

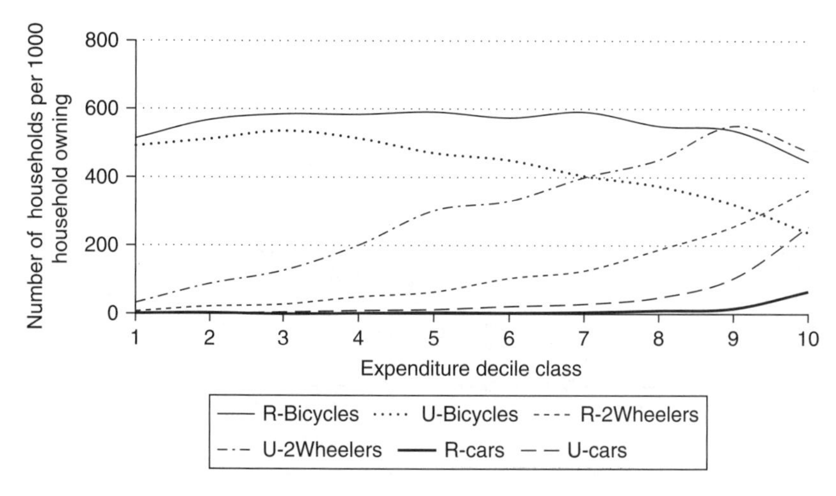

Source: NSSO (2012).

Figure 11.9 Vehicle ownership by expenditure decile class

wheelers than a two-wheeler; therefore, there is a drop in the two-wheeler ownership level for this class. Only a small share of the richest class (6.6 per cent) in rural areas own a car while an increasing trend in car ownership starts from the seventh decile class and a rapid growth is visible from the ninth decile class. In urban areas, 25 per cent of the richest class possess a car.

Interesting insights are also obtained by looking at the relationship between income distribution and ownership of other consumer durable goods (Figure 11.10). The difference between urban and rural living becomes very clear here. For example, whereas 43 per cent and 66 per cent of lower-income class households own television sets and electric fans, respectively in urban areas, the share falls to 10 per cent and 20 per cent, respectively in rural areas. But as income increases, the ownership level increases proportionately in rural areas whereas in urban areas it reaches a saturation level beyond the fourth expenditure decile. While durable goods like air conditioners, washing machines and refrigerators are used only by the richest segment of the rural population, these appliances are becoming more popular with urban consumers and an increasing level of ownership of such appliances is found as income grows. This implies that an urban consumer is likely to lead a more energy-intensive life as income increases.

A comparison of the recent survey with previous surveys indicates a rapid progression in ownership since the early 1990s, with a particularly

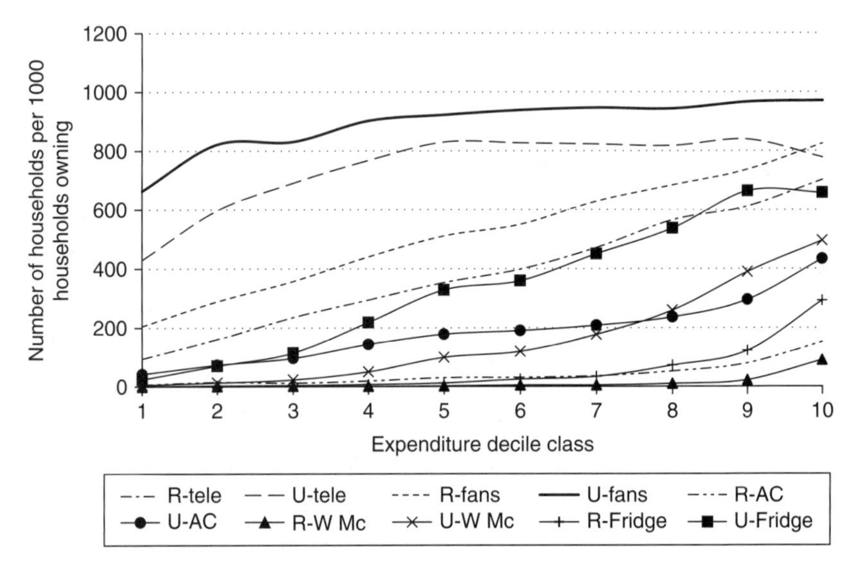

Source: NSSO (2012).

Figure 11.10 Ownership pattern of electric appliances by expenditure decile class

steep increase since 2004–05 in the rural areas (Figure 11.11). Bicycle, television and electric fan ownership have grown rapidly over the 16-year period whereas the holding of other appliances and ownership of two-wheelers or cars is catching up. This has significant implications for future energy demand in the country.

The trend of durable goods ownership in urban areas is quite different from that of rural areas (Figure 11.12). Bicycle ownership has reached saturation level and is showing a declining trend. Television and electric fan ownership is also reaching saturation level but a faster growth in consumer appliances such as refrigerators and air conditioners and in two-wheeler ownership is clearly visible. Car ownership has shown the least growth so far.

ECONOMIC GROWTH AND INCOME DISTRIBUTION

Although the Indian economy traditionally recorded an average annual growth close to 4 per cent (the so-called Hindu growth rate) prior to

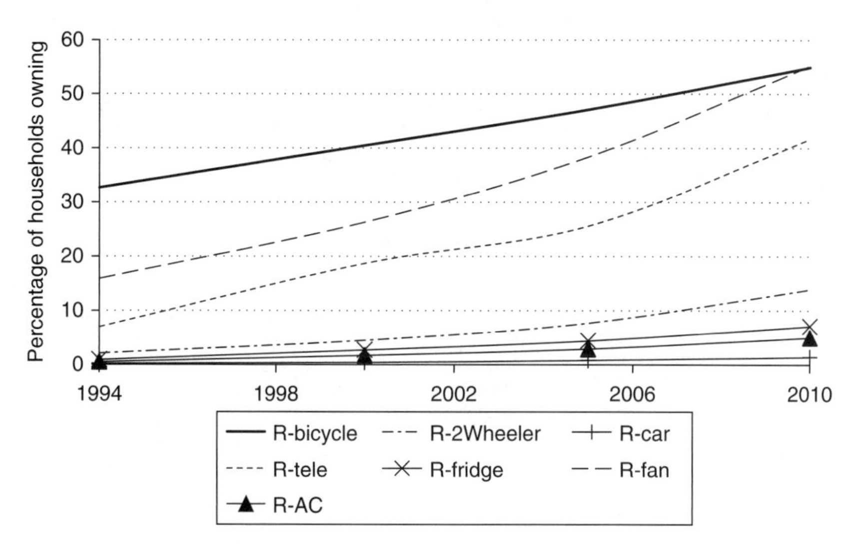

Source: NSSO (2012).

Figure 11.11 Trend of durable goods ownership in rural India

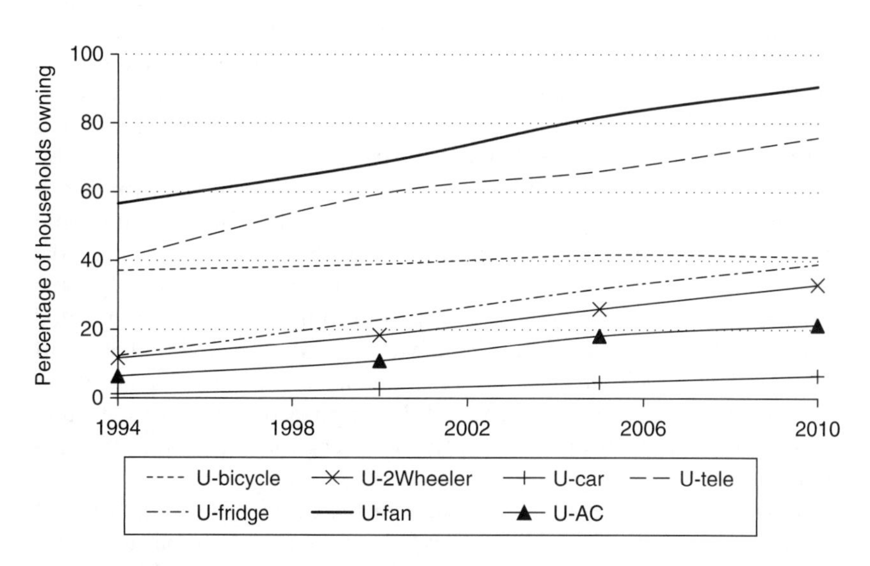

Source: NSSO (2012).

Figure 11.12 Trend of durable goods ownership in urban India

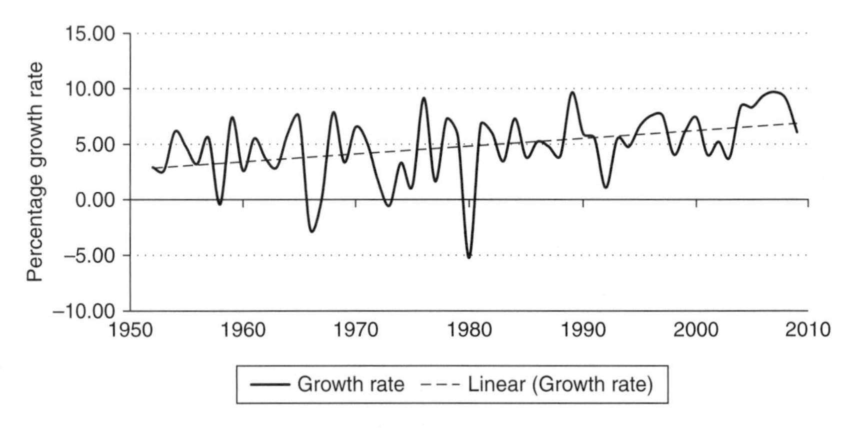

Source: RBI (2011).

Figure 11.13 India's economic growth

the 1980s, the country has seen economic improvements since 1991 (Figure 11.13). The growth in economic activities has been particularly strong since the new millennium and even in the period of global economic recession and economic distress, the Indian economy has been reported to maintain an above-average performance. Liberalization of the economy in the 1990s and 2000s, and consequent growth in investment, has improved the economic climate. The performance of the manufacturing sector is improving and the interest burden arising from public sector borrowing is falling with respect to GDP (Kharas, 2010).

Despite economic growth, poverty remains a main issue in India due to poor income distribution (Figure 11.14). According to data from PovcalNet, 34 per cent of the rural population and 29 per cent of the urban population lived below the poverty line in 2009. Although the government claims that poverty is falling in India, this has been debated by others. Yet it is clear is that the urban population is slightly better off compared with their rural counterpart, and a shift towards the right as a result of higher economic growth will bring poverty down in the future. This has the prospects of significantly increasing the middle-income class in the future.

Given all of these trends, it is clear that the country is on the brink of significant changes, many of which will have significant implications for the country's energy and environmental future.

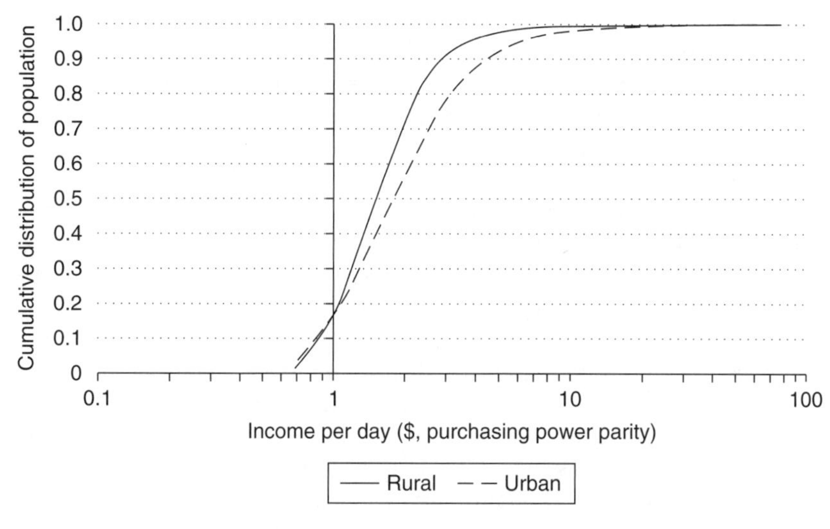

Source: PovcalNet, the online tool for poverty measurement developed by the Development Research Group of the World Bank, http://iresearch.worldbank.org/PovcalNet.

Figure 11.14 Income distribution in India in 2010

ANALYTICAL FRAMEWORK: CHOICE AND DESCRIPTION

To analyse the effect of demographic change, economic growth and lifestyle changes on future residential energy demand in India, this chapter presents a simplified analysis using an end-use framework. However, before elaborating the methodology used here it is important to know how other researchers have addressed similar issues. This also provides a better justification for the work presented. Accordingly, a brief review of the relevant literature is presented, followed by the elaboration of the methodological framework used.

Literature Review

As mentioned above, a number of studies have analysed the effect of changing economic structure and lifestyles on India's energy demand. Pachauri and Spreng (2002) used an input-output based approach to analyse and compare the direct and indirect energy demand over a ten-year period (1983/84–1993/94). They combined the household expenditure survey data with energy intensity information to better understand the implications of household energy demand on the overall primary energy demand of the country. They found that the direct and indirect energy requirements of Indian households contribute more than 75 per cent of total primary energy use in the country. While the study produced valuable insights, it focused on a representative Indian consumer without capturing the rural–urban dichotomy or income distribution aspects.

Filippini and Pachauri (2004) estimated the elasticities of electricity demand in urban households using a household expenditure survey for 1993–94. The study analysed the differences in electricity elasticity in three seasons of the year (winter, summer and monsoon) and used dummy variables to consider the influence of geographical location (regional dummies). The authors found that their own price elasticities were lower than other studies while the income elasticities were similar. They concluded that high income elasticity of electricity demand in urban areas signals a rise in electricity demand as the economy develops.

Pachauri (2004) used the same survey information indicated above but extended the econometric analysis to entire household energy demand (as opposed to just electricity as in Filippini and Pachauri, 2004). The study found that the most significant factor influencing energy demand was household expenditure per capita, with dwelling attributes, family characteristics and demographics being important to some extent.

Pachauri and Jiang (2008) present a comparative analysis of household energy transition in China and India using a set of three indicators, namely changes in the quantities of energy used, changes in the percentage of persons using different types of energy and the shifting pattern and structure of household energy consumption. This descriptive study relies on household surveys in both the countries and found that urban households consume more commercial energies than their rural counterparts in both the countries but more total household energy is used in the rural areas due to high reliance on inefficient fuels. Ekholm et al. (2010) used a microeconomic choice model that utilized the household survey of 1999–2000. The paper focuses mainly on cooking energy choices and analyses the effects of policy in improving the penetration of modern cooking fuels among the poor.

In contrast to the above micro-level analysis, the World Bank (2008) presents a detailed study of residential electricity demand for India in 2031–32 using the end-use approach. It used a four-step approach to forecasting electricity demand. First, it projected the number of households in urban and rural areas and their expenditure on an annual basis relying on a household expenditure survey of 2004–05. Second, the number of electrified households was estimated using demographic information. Third, the appliance ownership was forecast using past appliance sales and ownership information. Fourth, electricity consumption was forecast by combining the above factors. The study developed two scenarios – one reference scenario and the other alternative scenario where the case of more efficient technologies was analysed. It found that although the economy is assumed to grow at 7.8 per cent per year on average, the electricity demand grows at 5.8 per cent per year due to lower income elasticity at the household level. The alternative scenario suggests a further reduction in electricity demand, although a part of this may be taken back due to the rebound effect.

Similarly, de la Rue du Can et al. (2009) used the end-use approach to analyse residential and transport energy demand in India in 2020. It used household survey data of 1999–2000 and found that by 2020, biomass will play a comparatively lower role while electricity consumption per household will quadruple between 2000 and 2020. Similarly, energy used by cars is likely to grow at an annual rate of 11 per cent during this period.

Bin and Dowlatabadi (2005) used the consumer lifestyle approach in the USA to analyse energy demand impact. Wei et al. (2007) and Feng et al. (2011) presented studies of direct and indirect effects of urban and rural consumers on China's energy demand using the same approach.

Although the above studies have analysed the issue, they are based on older data. As more recent information is available, an attempt is made to present a simple impact analysis using the end-use approach.

Description of Methodology

The analysis carried out in the present chapter follows the end-use (or bottom-up) approach and is described below. Energy demand is essentially estimated by considering the relevant population in a given category and its per capita energy consumption. Energy consumption information for the urban and rural population per income class (in per capita terms in physical units) is captured from the household survey reported in NSSO (2012). The forecast for the urban and rural population is taken from *World Population Prospects 2010* and *World Urbanisation Prospects 2009* revision. This forms the first step.

As the energy demand pattern varies by income class, this analysis considers the distribution of the urban and rural population by income class as well. This is obtained using the Lorenz Curve and the parametric estimation was done using PovCal software of the World Bank. Future distribution of the urban and rural population by income class assumes that the same distribution as in 2010 continues. We assume that the economy grows at an average rate of 7 per cent per year for the entire period of study (2010–30), which directly improves the per capita income proportionately. We adjust the economic growth rate by the population growth rate to arrive at the effective growth rate of per capita income. Following Kharas (2010), we use three income groups – poor, medium income and rich separately for the urban and rural population. We assume like Kharas (2010) that a household of four members with a daily income below $10 (in purchasing power parity terms) is poor, while those falling within $10 and $100 are middle-income households. Any household with an income above $100 per day is considered as rich.

Energy consumption by fuel type per person per month for each income class is taken from the NSSO (2012) survey. The fuel use pattern for each decile class in urban and rural areas is first considered. This information is then aggregated into three categories to reflect the poor, medium-income and high-income categories. The tenth decile is considered as the high-income group for both areas. For the rural poor category, an average of the first seven decile classes is considered while for the urban poor category, the first six decile classes are considered. The remaining classes are considered as middle-income groups. The results obtained for the reference case in 2010 are calibrated and reconciled against 2009 energy consumption data reported by the International Energy Agency IEA. As the survey was carried out in 2009–10, the results should be compared against the relevant period of 2009 and 2010 but this was not possible. As 2009 data are the most recent data reported by the IEA, we have used this information to calibrate and adjust the model.

Fuel-wise energy consumption is obtained by multiplying consumption per person with the relevant population size. We considered three scenarios for forecasting: (1) the reference scenario assumes that per person consumption in 2010 remains unchanged in future years; this captures the effects of demographic transition and the urbanization effect; (2) a low energy demand growth scenario is considered where per person consumption pattern changes slowly at an annual growth rate of 2 per cent for the entire period; (3) a high demand growth scenario following an assumed growth path as follows: (a) for low- and middle-income groups, the consumption pattern follows the pattern of the highest decile group by 2020; the demand per person grows at 3 per cent per year for electricity and LPG while firewood, kerosene and coal demand declines by 1 per cent per year; (b) for the high-income category, a growth rate of 3 per cent per year is considered for electricity and LPG consumption compared to 2010. For urban consumers, coal and firewood consumption is assumed to decline to zero over the study horizon and minimal kerosene consumption for stand-by use is assumed. For rural consumers, a 1 per cent decline rate is used for firewood and coal while kerosene consumption is assumed to remain unchanged.

The analysis was done for a period of 20 years (until 2030) taking a five-year snapshot. Although the analysis follows a simple framework, it provides useful insights because of the disaggregated level of analysis that captures the urban–rural differences in energy demand, fuel mix diversity and the shift along the energy ladder of a developing economy.

ANALYSIS OF RESULTS

The expected distribution of population by income decile is presented in Figure 11.15. As indicated above, the distribution does not change as such but shifts rightwards as income improves as a result of economic growth. As a consequence, the size of the Indian poor category dramatically reduces and the middle-income group swells exponentially. By 2030, 95 per cent of the urban population and 93 per cent of the rural population are likely to enter the middle-income category. The size of the Indian rich category is likely to reach 4 per cent of the urban population (or about 23 million) by 2030. These demographic changes will have major implications for the energy needs of the country.

Although the analysis was done for each fuel using original physical units, the results are presented below in a common unit, million tonnes of oil equivalent (Mtoe) for ease of comparison. It is important to note that using the population distribution estimated above and the survey-based per capita energy demand for each fuel resulted in some differences in the total energy demanded in 2010, the analysis produced a higher estimate of electricity demand while other fuel demands were underestimated. To realign the results to the IEA data, a correction factor was introduced for each fuel.

The total residential energy demand in India under three different scenarios is presented in Figure 11.16. It shows the commercial energy demand and the demand including firewood.[1] It can be seen that in the reference scenario, the commercial energy demand is likely to increase to 78 Mtoe by 2030 from 38.8 Mtoe in 2010. Thus, assuming that current energy patterns remain unchanged going forward, the demographic effect and economic growth of the country alone would require an additional 39 Mtoe for the residential sector. If the demand for firewood is added, the demand growth rate becomes moderate as the firewood demand grows at a much slower pace. It needs to be mentioned that in the reference scenario, modern energies play a minority share in 2030, while in the low and high scenarios, the situation reverses and the firewood share drops to 40 per cent in 2030.

The picture changes substantially in two alternative scenarios. While

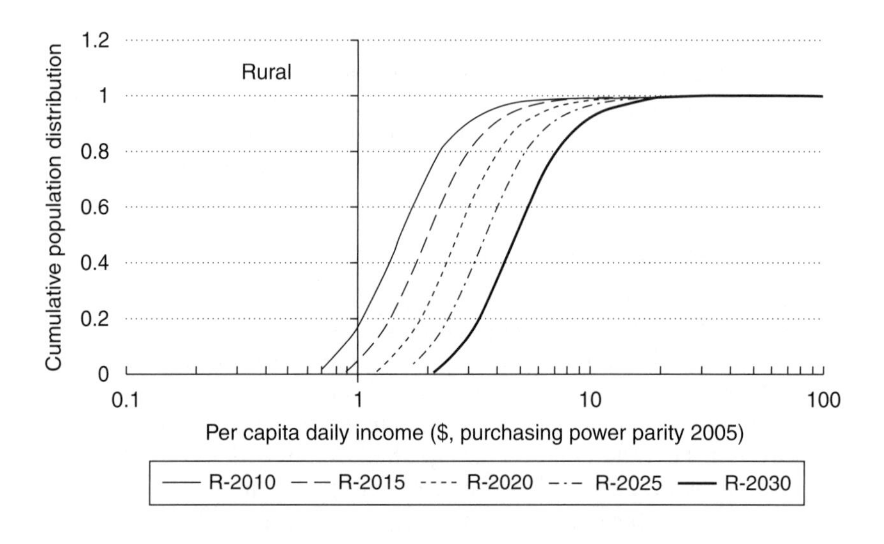

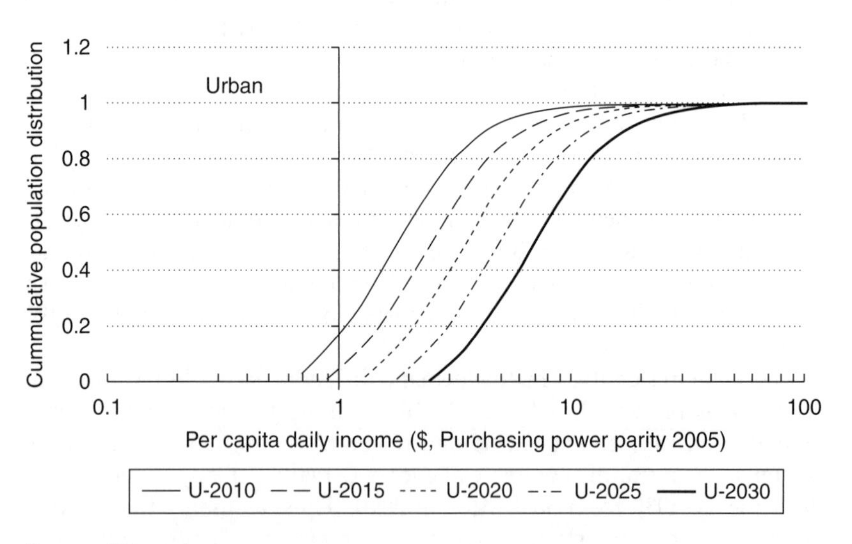

Source: This study.

Figure 11.15 Income distribution of Indian population up to 2030

the commercial energy demand increases faster in the high scenario, thereby trebling the demand by 2030 compared to 2010, the firewood demand peaks in the low scenario as it is assumed that the population continues to use firewood alongside commercial fuels. Thus, the total

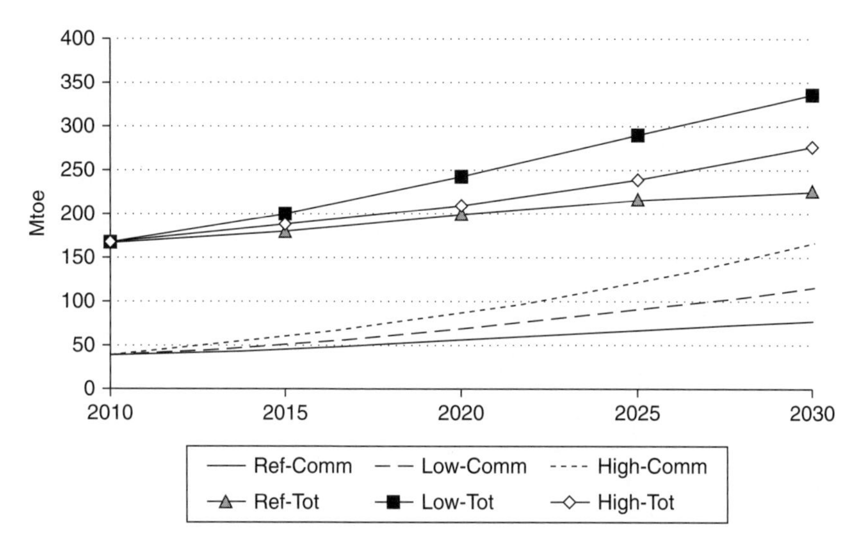

Source: This study.

Figure 11.16 Residential energy demand outlook for India

energy demand including firewood is highest in the low scenario. This presents an interesting outcome – economic growth and demographic transition are likely to accentuate the demand for modern energies in the residential sector. If the population imitates the energy consumption style of its higher-income peers, India is likely to see a rapid growth in its commercial energy demand. While such a change may impose higher import burden on the economy, it is likely to reduce the external costs associated with firewood use and benefit the poorer section of the population immensely.

In terms of fuels, similar interesting insights are obtained looking at Figure 11.17. Both electricity and LPG demand increases more than 2.5 times by 2030 in the reference scenario. In fact, these two forms of energy drive the demand for commercial energies in the residential sector. The demographic transition and economic growth will push up electricity and LPG consumption significantly. However, India has a relatively high electricity consumption to primary energy input ratio (4:2 in 2005 according to de la Rue du Can et al., 2009) due to a relatively low conversion efficiency of electricity generation and high transmission distribution loss. This implies that a rapid growth in electricity demand also requires a fourfold increase in electricity generation, which can easily be seen as a source for a problem.

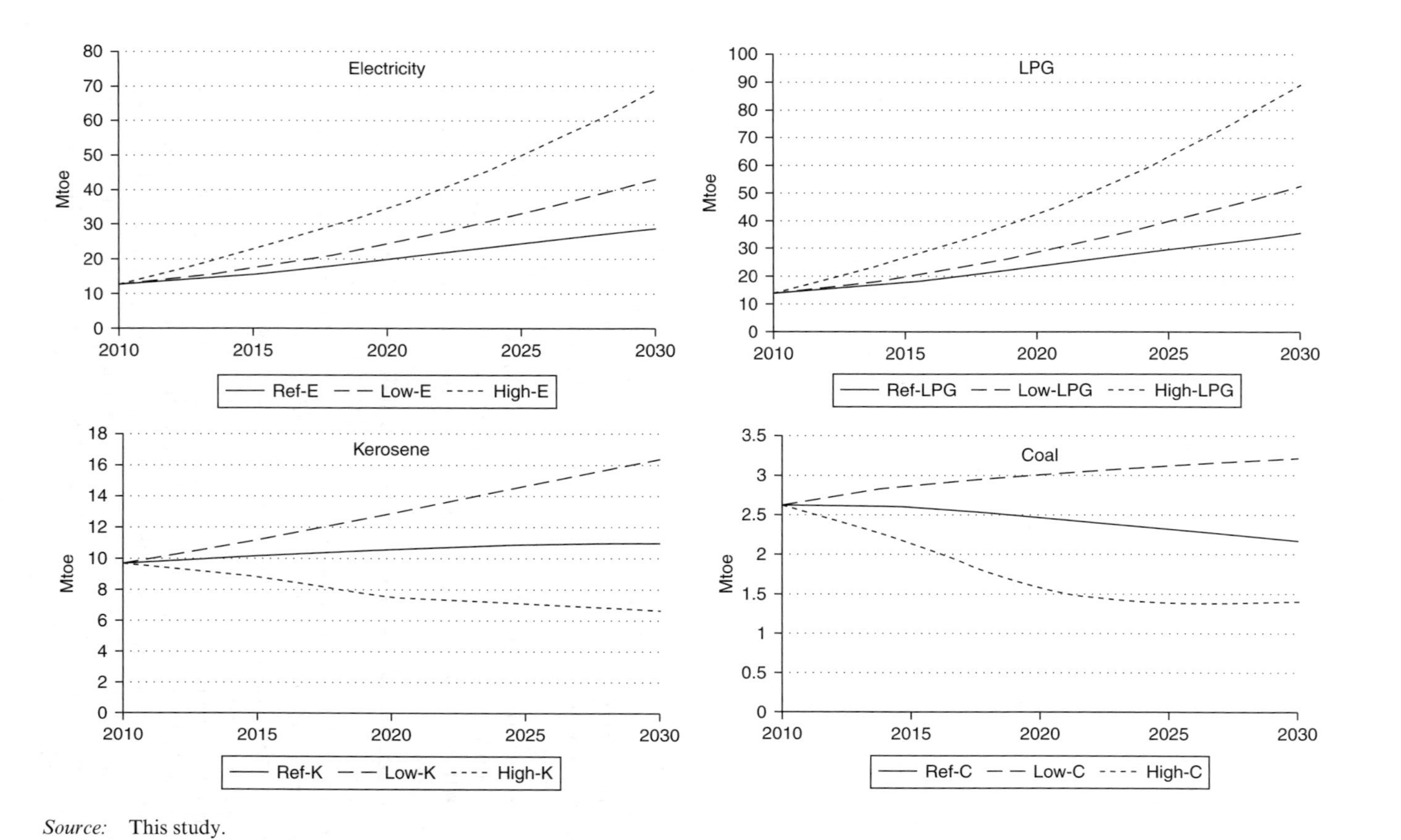

Source: This study.

Figure 11.17 Indian residential energy demand

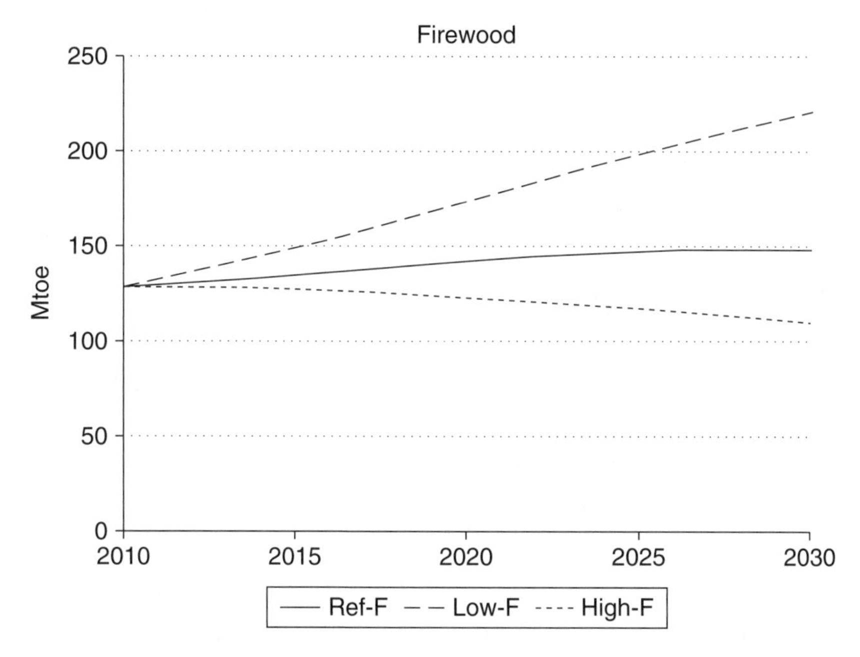

Source: This study.

Figure 11.18 Firewood demand in India

Kerosene and coal, on the other hand, are likely to follow a different development path. In the reference case, kerosene demand marginally grows while coal demand marginally falls. Both fuels see a dramatic decline of their demand in the high scenario as electricity and LPG displace them. On the other hand, in the low scenario, their demand increases. The demand for firewood also follows a similar path (Figure 11.18). The demand grows marginally in the reference scenario but increases in the low scenario while falling rapidly in the high scenario. In fact, firewood demand in the high scenario in 2030 will be about 50 per cent of the low scenario due to substitution of woody fuel.

It is also interesting to note that modern energy demand is driven by the urban population (Figure 11.19) and this continues until 2030, despite the growing middle class in the rural areas. The urban share of electricity is expected to increase to 70 per cent by 2030 from 60 per cent in 2010 in the reference case. In the high scenario, the urban share of electricity demand increases even further to 80 per cent. The share of LPG remains even higher at above 80 per cent. On the other hand, rural areas dominate in coal, firewood and kerosene.

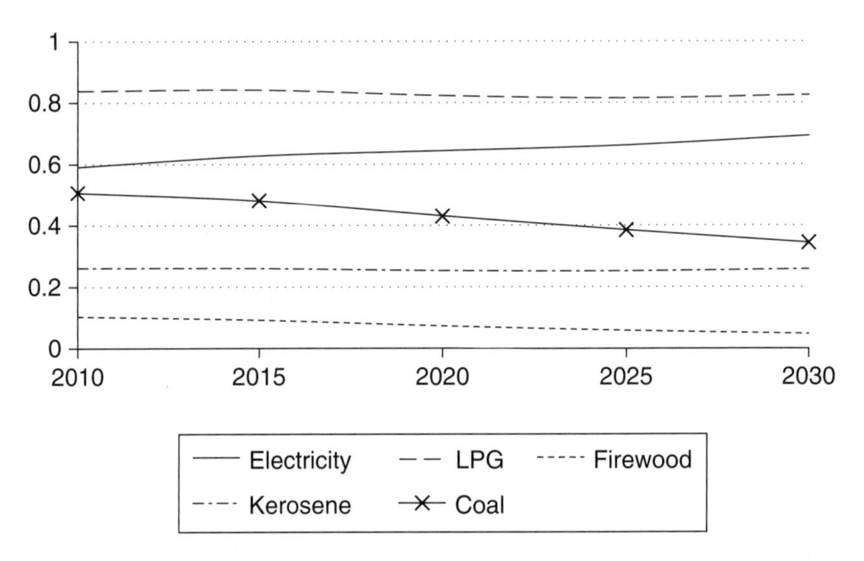

Source: This study.

Figure 11.19 Urban share in the residential energy demand (reference case)

Finally, the energy demand by income class reveals a clear transition. A high share of energy demand (including firewood) from the Indian poor category declines over time and the share of the middle-income group increases. By 2030, the middle-income group will completely dominate the energy demand (Figure 11.20). The rise in the number of the middle-income population is responsible for such a transition. The situation is slightly different for commercial energies where the share of the high-income group increases slightly over time as well but the middle-income group continues to dominate in all scenarios.

In summary, the demographic changes, economic growth and consequent lifestyle changes will bring profound changes in the residential energy consumption in India over the next 20 years. The demand for commercial energies will rise rapidly and the emergence of a large middle-income group is likely to drive energy demand. Rapid urbanization of the country is also likely to increase the commercial energy demand from urban consumers. As electricity and LPG are likely to become the preferred fuels of the urbanized population, India will need a rapid expansion of its supply infrastructure to cope with such a transition.

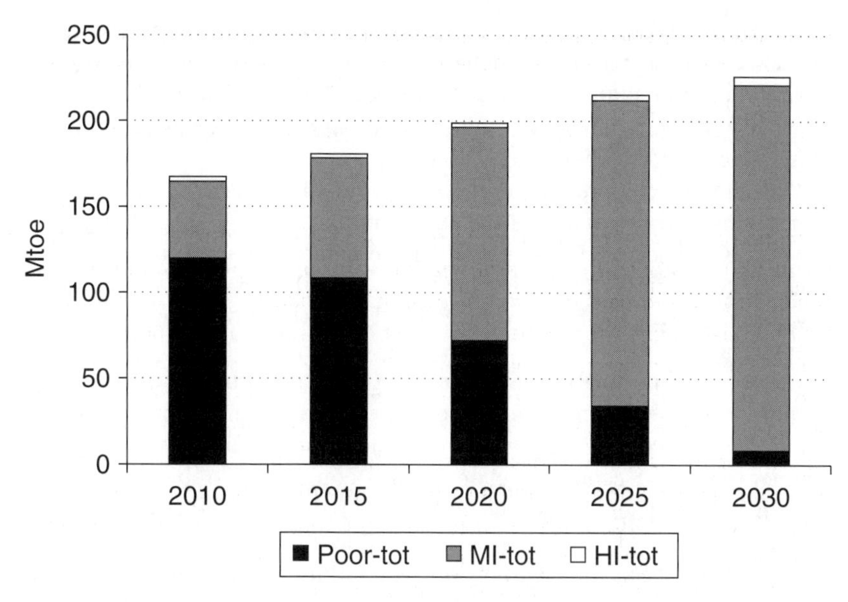

Source: This study.

Figure 11.20 India's residential energy demand outlook by income class

WAY FORWARD AND CONCLUSIONS

Clearly, there is huge potential for improving the quality of life without necessarily following an energy-intensive lifestyle. As the economy transforms, there is a window of opportunity to follow an alternative development path that relies more on sustainable energy futures. This has been elaborated in Bhattacharyya (2010) and is briefly recapitulated below. India would have to consider a strategy with the following features to ensure a transition to a low energy intensity economy: (1) manage energy demand carefully; (2) adopt internationally best-practice technologies; (3) follow good governance and modern management practices; (4) use indigenous resources effectively; and (5) ensure access to clean energies by deploying decentralized and distributed supply options.

The residential sector offers a huge potential to manage the demand carefully through pricing signals and demand-management policies. While a number of initiatives have been taken in the right direction, the potential remains largely untapped. As the country is developing long-life assets, it is important to ensure that international best practices are used to maximize efficient use of energy and optimize clean energy use in the residential sector. Similarly, the reduction in system losses through good governance and management practices can also reduce fuel needs for electricity conversion. Unless these aspects are given adequate attention immediately, future challenges will multiply.

This chapter has presented an overview of the transformation India is undergoing and has analysed the energy implications of such changes for the residential sector. It has indicated the demographic transition taking place in the country and combined this with the prospect of economic development over the next two decades. Using the end-use approach, the chapter has provided India's future energy demand outlook for the residential sector. The population dividend along with the prospect of continued economic growth has the potential for increasing residential energy demand manifold. The commercial energy demand could double just because of demographic and economic changes and could even treble in the next 20-year period. There is also the prospect of reducing reliance on firewood as the size of the Indian poor population reduces. The middle-income group is likely to emerge as the major energy demanding group in the country and the urban consumers will demand more energy than their rural counterparts. Clearly, as a nation in transition India has a window of opportunity to shape an energy-efficient future and thereby ensure a sustainable energy future. But this can only be achieved through a careful strategy.

The analysis presented here has a number of limitations. The appliance stock has not been captured in the analysis. Similarly, the impact of economic development has been taken indirectly through a shifting consumer class but it is possible to capture economic drivers directly in the analysis. Moreover, the analysis was based at an all-India level and did not consider the differences existing at the state level. Further research can focus on these issues.

NOTE

1. By commercial energy we mean energies that are sold in the market and bear a price. These are coal, LPG, electricity and kerosene in our case. Although firewood enters the market in many places, it is still collected and used.

REFERENCES

Bhattacharyya, S.C. (2010), 'Shaping a sustainable energy future for India: management challenges', *Energy Policy*, **38** (8), 4173–85.

Bin, S. and H. Dowlatabadi (2005), 'Consumer lifestyle approach to US energy use and the related CO2 emissions', *Energy Policy*, **33** (2), 197–208.
Bloom, D.E. (2011), 'Population dynamics in India and implications for economic growth', Program on the Global Demography of Aging, Harvard School of Public Health working paper 65, Harvard University, Cambridge, Massachusetts.
De la Rue du Can, S., V. Letschert, M. McNeil, N. Zhou and J. Sathaye (2009), *Residential and Transport Energy Use in India: Past Trend and Future Outlook*, Berkeley, CA: Lawrence Berkeley National Laboratory, University of California.
Ekholm, T., V. Krey, S. Pachauri and K. Riahi (2010), 'Determinants of household energy consumption in India', *Energy Policy*, **38** (10), 5696–707.
Feng, Z.H., L.L. Zou and Y.M. Wei (2011), 'The impact of household consumption on energy use and CO2 emissions in China', *Energy*, **36** (1), 656–70.
Filippini, M. and S. Pachauri (2004), 'Elasticities of electricity demand in urban Indian households', *Energy Policy*, **32** (3), 429–36.
Government of India (2012), *Provisional Population totals, Paper 1 of 2011 India, Series 1, Chapter 3: Size, Growth Rate and Distribution of Population*, available at www.censusindia.gov.in/2011-prov-results/prov_results_paper1_india.html (accessed 18 December 2012).
Kharas, H. (2010), 'The emerging middle class in developing countries', OECD working paper 284, OECD Development Centre, Paris.
MGI (McKinsey Global Institute) (2010), *India's Urban Awakening: Building Inclusive Cities, Sustaining Economic Growth*, available at www.mckinsey.com/insights/urbanization/urban_awakening_in_india (accessed June 2013).
MGI (McKinsey Global Institute) (2012), *Urban world: Cities and the Rise of the Consuming Class*, available at MGI-Urban-World_Executive_Summary_June_2012[1] pdf (accessed 19 December 2012).
NSSO (National Sample Survey Office) (2012), available at www.indiaenvironmentportal.org.in/category/publisher/national-sample-survey-office (accessed 19 December 2012).
Pachauri, S. (2004), 'An analysis of cross-sectional variations in total household energy requirements in India using micro survey data', *Energy Policy*, **32** (15), 1723–35.
Pachauri, S. and D. Spreng (2002), 'Direct and indirect energy requirements of households in India', *Energy Policy*, **30** (6), 511–23.
RBI (2011), *Handbook of Statistics*, Mumbai: Reserve Bank of India.
Singh, S.K. and K.S. Hari (2011), 'International migration, remittances and its macroeconomic impact on Indian economy', Indian Institute of Management working paper 2011–01–06, Ahmedabad, India.
United Nations (2010), *World Urbanisation Prospects: The 2009 Revision, Highlights*, New York: United Nations.
United Nations (2011), *World Population Prospects – the 2010 Revision, Volume 1 Comprehensive Tables*, available at http://esa.un.org/unpd/wpp/Documentation/pdf/WPP2010_Volume-I_Comprehensive-Tables.pdf (accessed 18 December 2012).
Wei, Y.M., L.C. Liu, Y. Fan and G. Wu (2007), 'The impact of lifestyle on energy use and CO2 emission: an empirical analysis of China's residents', *Energy Policy*, **35** (1), 247–57.
World Bank (2008), 'Residential consumption of electricity in India: documentation

of data and methodology', in background paper, *India: Strategies for Low Carbon Growth*, Washington, DC: The World Bank.

World Bank (2010), *Migration and Remittances Factbook, 2011*, Washington, DC: The World Bank.

12. Institutional and community-based initiatives in energy planning

Steven M. Hoffman and Angela High-Pippert

INTRODUCTION

On 4 May 2007, Greensburg was a declining farm community in south-central Kansas with a population of about 1400. That evening, an EF-5 strength tornado touched down more than 75 times, killing 11 people and destroying or severely damaging 90 per cent of the city. The storm left a trail of debris more than 22 miles long and 1.5 miles wide. The initial response by the community and its municipal utility was focused on the restoration of the existing electrical system (Billman, 2009, p. 5). Once attending to these immediate needs, however, the community began to envision a very different future, one based on an understanding that the nearly complete destruction of the town presented 'a unique opportunity to create a strong community devoted to family, fostering business, [and] working together for future generations' (Billman, 2009, p. 5). Such a community could, in large part, be built upon an energy future that would take advantage of 'Greensburg's vast wind resources . . . reduce energy use in buildings, industry, and infrastructure, use renewable sources for electricity and heat at the community and distributed scales, use alternative transportation vehicles, fuels, and infrastructure, and support new approaches with institutional and administrative actions' (Billman, 2009, p. 12). Four years on, the town's continuing commitment towards an alternative energy future was evidenced by a new school, hospital, city hall, arts centre and a whole new downtown business district, all built to be energy efficient and tornado resistant, which, at least according to numerous news accounts, have turned 'Greensburg into one of the "greenest" and most tornado-proof places in the country' (Bowers, 2011).

The desire to fashion a more locally oriented and renewably based energy future is not limited to those places presented with a blank slate occasioned by a natural disaster. The city of Boulder, CO, for instance, is one of more than 100 US cities that have recently entertained the concept of municipalization. In Boulder's case, this would entail condemning the assets owned by the city's current service provider, Xcel Energy, and then using these assets to create a local, or municipally owned, distribution and/or generation utility. The proposal is based, in part, on the city's dissatisfaction with Xcel's current generation mix, which 'is tied to several large coal plants until the 2030s' (Silverstein, 2011). In contrast, a municipal utility would, according to its supporters, offer 'greater flexibility and local control, the ability to reduce local carbon emissions, more opportunities to target investments in a manner consistent with local priorities, and long-term economic benefits through the creation of a more competitive industry and greater energy innovation' (City of Boulder, 2011).[1]

While reconstruction and municipalization are both long, tedious and, in the case of the latter, generally unsuccessful efforts,[2] they nonetheless speak to a growing recognition that effective climate mitigation strategies will require an equal mix of large-scale, centralized production technologies and decentralized or distributed forms of generation that are locally produced and/or consumed (California Energy Commission, 2001; O'Brien, 2009). The latter, in turn, will necessitate action at the local and community levels and the involvement of actors that can stimulate significant behavioural changes on the part of both households and local businesses (*UK Low Carbon Transition Plan: National Strategy for Climate and Energy*, 2009). At the present time, however, the strongest appeal of so-called 'community energy' initiatives has been as a rhetorical device useful for capturing public support for otherwise controversial electrical projects, notwithstanding the heroic efforts of places like Greensburg (Hoffman and High-Pippert, 2005b, 2007). To move beyond community as rhetoric, however, raises the question posed by Amory Lovins more than three decades ago, namely, whether or not it is possible for 'everyone [to] get into the act, unimpeded by centralized bureaucracies [and make] energy choices through the democratic political process' (1977, p. 99). In other words, is it possible to embed a meaningful and substantive idea of 'community' when doing 'community energy' programmes? In order to answer this question, it is useful to begin with a clear understanding of the many competing notions that currently surround the term.

COMMUNITY ENERGY: A CONCEPTUAL FRAMEWORK

As illustrated in the cases of Greensburg and Boulder, and as shown in Figure 12.1, community energy is a highly contested notion that carries with it a multitude of meanings, one of the most important being the

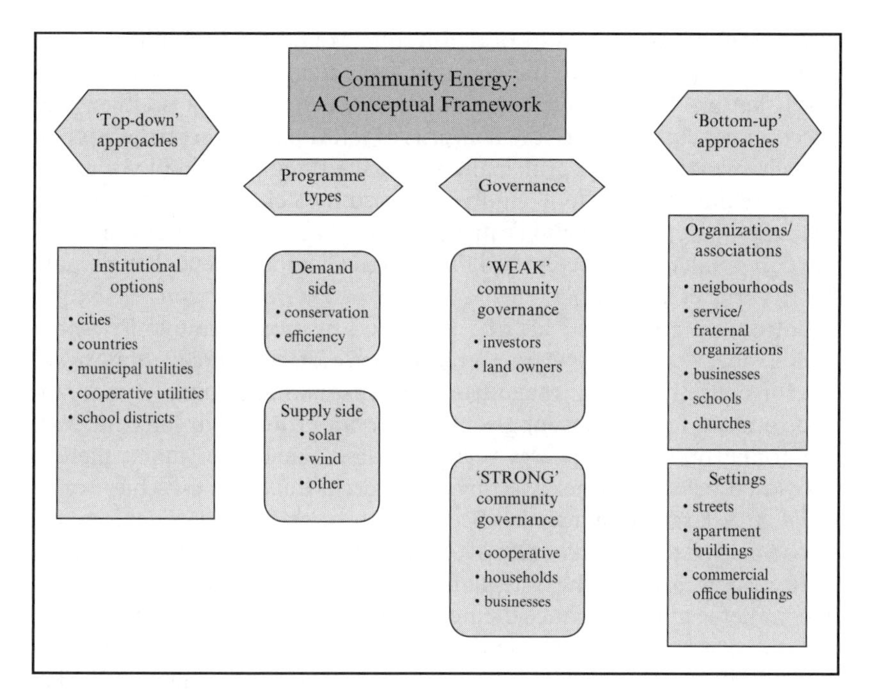

Figure 12.1 Community energy: a conceptual framework

distinction between 'top-down' or institutionally driven efforts versus 'bottom-up' initiatives. The former would include programmes that are developed and delivered by a variety of institutional alternatives, including cities, counties, municipal and cooperative utilities and school districts.

'Bottom-up' initiatives, on the other hand, refer to those that develop independently of an institutional sponsor. Such initiatives might develop whole cloth, arising from acts as unremarkable as a casual conversation between neighbours or friends concerned about a vaguely energy-related concern, that is, a proposed new energy-related facility such as a new or expanded transmission line, substation or generating station, or even a generic interest in reducing their electricity bills or, more amorphously, their carbon footprint (Hoffman and High-Pippert, 2010). In this respect, various informal organizations such as neighbourhood block clubs, local reading or garden clubs, groups of stay-at-home mothers or fathers and so on might well provide the sort of informal meeting ground necessary for the beginnings of an energy-related initiative.

More formal organizations, such as churches, schools, scout troops and so on, or what Kretzmann and McKnight (1993) refer to as social or community assets, can also be a fertile ground for community-based energy initiatives, despite the fact that such civic associations have missions that are far removed from questions of how energy is either produced or consumed. Thus, the concerns of a few parishioners, the passion of a charismatic school principal or even the daily pestering by a group of highly motivated school children inspired by a classroom activity who goad their parents into action can all serve as the starting point for innovative community-based actions (Galluzzo and Osterberg, 2006; Peters et al., 2011; Verba et al., 1995).

Bottom-up efforts can also be generated by other community actors such as local business owners. As in the case of households, opportunities for collective action range from the informal and initially unorganized to more formal and long-standing organizations that have, at best, a marginal interest in energy-related issues. The former might include individual business owners occupying different buildings on a busy retail street or a large commercial office building with multiple small tenants whose disparate interests unite over a shared interest in reducing heating or electricity costs. On the other hand, a few members of the local chamber of commerce might convince their chapter to devote a monthly meeting to issues of energy efficiency and in doing so stimulate longer-term efforts at 'green building design', Leadership in Energy and Environmental Design (LEED) certification or other energy-related programming.

In both cases, whether focused on householders or business groups, a distinction can be made between social organizations and the physical settings in which these organizations can arise. Thus, a 'neighbourhood' could be a high-density Jane Jacobs-like (1961) aggregation of city streets, a suburban housing development, a single apartment or office building, or a senior living complex. In the case of businesses, an association might be geographically defined, that is, a single office building or by a common business interest such as in the sale of a particular product.

A second important distinction illustrated in Figure 12.1 are the two basic types of programmes common to the energy sector, namely, supply-side programmes, the most common being wind projects though there is increasing attention being given to solar projects, and demand-side initiatives that focus on behavioural changes at the level of the individual household or business. As shown in the case of both Greensburg and Boulder, there are multiple ways to combine the locus of activity, or top-down versus bottom-up programmes, with both demand- and supply-side options. For instance, certain types of projects, such as the development of large-scale wind farms, would seem to be best suited to well-funded institutional actors working with local landowners and private investors while others, such as technology swaps, that is, convincing people to switch to high-efficiency light bulbs, might work best in social or physical settings requiring little in the way of capital, time or effort on the part of individuals or firms.

A third important aspect of community energy programming concerns the sort of governance that might characterize any programme or initiative, the first step in which is identifying those parties responsible for defining an initiative's essential features. Another way of asking this question is who 'owns' the project, that is, does the ultimate ability to define its scope and nature reside within the community or social organization or does it reside some distance from a significant number of community members? Figure 12.1 identifies two potential governance structures: a 'weak' versus a 'strong' form of community governance. In the former, the reach of community members is constrained, perhaps being limited to financial partners or, as is the case of most so-called 'community wind' projects, farmers fortunate enough to own the land upon which utility-scale wind machines can be built.

A 'strong' version of community governance, on the other hand, involves a much greater depth of community participation. In this case, many more households and/or businesses are assumed to be actively engaged in both the formative and operational stages of the initiative; just how much participation is required, however, is open to question. It might be the case, for instance, that a high level of participation is limited to a few very active individuals whose role it is to model desirable behaviours or to encourage their neighbours to engage in activities that require little time, effort or financial investment (Hoffman and High-Pippert, 2010). Whatever the level of participation, however, a strong form of governance would vest in the community the ability to define and exercise control over the most essential features of the project. This would be true even if its implementation is handed over to a local government authority that possesses the resources necessary to actually carry out the project.

'TOP-DOWN' OR 'BOTTOM-UP' INITIATIVES: WHICH WAY TO GO?

However intriguing might be the political and social implications of community energy, there exists a more immediate and practical question, namely, which type of programme will most effectively stimulate the behavioural changes on the part of both households and businesses required to achieve concrete emissions reductions?[3] Two bodies of literature are most useful in this regard: (1) that relevant to innovation and the diffusion of that innovation and (2) recent research in the field of social psychology related to how social norms encourage behavioural change.

Innovation and Change

The problem of innovation and its diffusion has occupied numerous academic literatures for at least the last several decades. While diverse in both scope and the questions of interest, much of this literature shares a common ancestry, namely, Everett Roger's (1995) work, *Diffusion of Innovation*, which defined an innovation as 'an idea, practice, or object that is perceived as new by an individual or other unit of adoption' (p. 11). Although Rogers's definition includes possibilities for individual-level analysis, most behavioural scientists have focused their attention on more system-level or macro-level analyses, rarely venturing into innovation at the community level or extremely decentralized or potentially diffuse units such as neighbourhoods, churches, senior centres, or small chambers of commerce.[4]

Walker's (1969) classic work on diffusion of policy innovations is representative of the literature. For Walker, and then Gray (1973), an innovation is defined as 'a program or policy which is new to the states adopting it, no matter how old the program may be or how many other states have adopted it' (Walker, 1969, p. 881). Following Walker's lead, the majority of political scientists have tended to use the terms 'innovation' and 'diffusion' to refer to the process by which ideas and policies spread from one state government to another (see Boushey, 2004; Emmert and Traut, 2003; Haider-Markel, 2001; Norberg-Bohm, 1999). Some of this research has emphasized the role of policy entrepreneurs, or individuals who, through an advocacy network, champion ideas until they become public policies (Baumgartner and Jones, 1993; Mintrom and Vergari, 1998). As Boushey writes, '[I]t seems clear that motivated individuals serve as the catalyst for implementing political ideas and implementing policy change' (2004, p. 6).

A more general review of the innovation and diffusion literature also reveals an emphasis on the 'practical'. According to Drazin and Schoonhoven (1996), 'the study of innovation appears to derive from practical rather than theoretical concerns [and] has been dominated by normative explanations of how to achieve an outcome seen as central to the interests of managers: increasing the number of innovations generated' (p. 1065). A key factor in explaining the resistance of firms and individuals to innovation is the presence of uncertainty; a process that erodes this uncertainty is therefore crucial to the adoption of innovation. Hence, the diffusion of innovation hinges on the extent to which various actions, perceptions, communication processes and sources, social norms and structures sufficiently reduce the potential adopter's uncertainty regarding the innovation. Rice, for instance, defines diffusion of innovation as 'the process through which an innovation (an idea, product, technology, process or service) spreads (more or less rapidly, in more or less the same form) through mass and digital media, interpersonal and network communication, over time, through a social system, with a wide variety of consequences (positive and negative)' (Rice 2009, p. 1).

These factors, that is, media, interpersonal communication, individualized social networks and the wider social system, interact in a variety of ways to affect the diffusion process. For instance, an individual in a social system may become aware of an innovation through various mass media. But a change agent, representing the institution or agency sponsoring the innovation, may increase the likelihood and diffusion of the adoption through interpersonal communication with the local opinion leader, who is densely connected with and influential for other members of the network. In other words, while mass media may be useful in the creation of knowledge, 'interpersonal channels are more effective in persuading an individual to adopt a new idea, especially if [they] are near-peers'. At the heart of the diffusion process, therefore, is the 'modelling and imitation by potential adopters of their network partners who have adopted previously' (Rogers, 1983, p. 18). Gerber and Green's (2000) field study of approximately 30000 registered voters comparing the effectiveness of various get-out-the-vote strategies, including personal canvassing, direct mail and telephone calls, comes to the same conclusion. As noted by Gerber and Green (2000), 'face-to-face interaction dramatically increases the chance that voters will go to the polls' (p. 661).

According to Weatherford (1982), the reason why 'near-peers' are so critical is that 'even if the content is discussed, for the receiver to adopt it and (potentially) to influence others requires not only comprehension of the message but also attitude change and commitment' (p. 122), a transformation that is more likely if information is communicated by peers rather than by experts (Rogers, 1983; see also Coleman, 1959; Weatherford, 1982). Leonard-Barton (1985) makes the same point, arguing that '[N] umerous marketing and diffusion studies have demonstrated that the more favorable information a potential adopter has received from peers, the more likely that individual is to adopt' (p. 914).

More specifically, one's location in a social system's communication network strongly affects the speed and extent of information/adoption diffusion (Valente, 2005). Weak ties, or infrequent communication with those who are not close, might well provide exposure to new ideas and information (Granovetter, 1974). However, frequent social or physical exposure to influence from salient others is especially crucial for reducing uncertainty about the innovation, fostering supportive social norms and persuading potential adopters. The importance of interpersonal communication networks in combination with targeted media messages on diffusion of innovations has been popularized through concepts such as the 'tipping point' or the need to influence a critical mass of adopters (Gladwell, 2000; Markus, 1987) and new product 'buzz' (Rosen, 2000). The recent development of online social networking websites are also said to enable rapid diffusion of information and persuasive norms (Wang et al., 2007).

The Power of Social Norms

The innovation and diffusion literature is not alone in suggesting the importance of community, peer-driven channels of change. Equally important is an emerging body of work that originates in the field of social psychology. Research on normative social influences clearly demonstrates the powerful effect of others' behaviour on our own, from classic works on the powerful effect of observing the development of social norms (Asch, 1951; Sherif, 1936) to more recent research that indicates that a written description of a social norm can be as powerful as direct observation in shaping conformity and individual behaviour (Parks et al., 2001). As Nolan et al. (2008) point out, this practice of providing a written description of peer behaviour, such as neighbours or members of reference groups, has been effective at both discouraging generally negative behaviours (for example, binge drinking) as well as encouraging more positive social and individual behaviours such as recycling. Goldstein et al. (2008), for instance, have assessed the effectiveness of messages used by hotels to encourage the reuse of towels. In this case, the standard approach of framing participation in the programme as protecting the environment was less effective than simply conveying the norm that most guests reuse their towels at least once during their stay.

Recent research in social psychology has focused on the connection between the influence of these sorts of so-called descriptive norms, which refer to how most people behave in a given situation, and an individual's willingness to engage in energy conservation behaviours. In their study of 810 Californians, Nolan et al. (2008) asked participants about their efforts to conserve energy as well as their reasons for doing so. The survey included a question designed to gauge the importance of social norms in this decision: 'In deciding to conserve energy, how important is it to you . . . (a) that using less energy saves money, (b) that it protects the environment, (c) that it benefits society, and (d) that a lot of other people are trying to conserve energy?' Interestingly, the results of this study indicate that although participants believed that the behaviour of their neighbours had the least impact on their own energy conservation, the reverse was actually true. Thus, Nolan et al. (2008, p. 921) explain that:

> Normative information spurred people to conserve more energy than any of the standard appeals that are often used to stimulate energy conservation, such as protecting the environment, being socially responsible, or even saving money. Descriptive norms had a powerful but underdetected effect on an important social behaviour: energy conservation.

The power of descriptive norms to motivate energy conservation is now being used to motivate residential energy conservation. For example, Xcel Energy, a Midwestern utility whose service territory stretches from the Canadian to the Mexican borders, is now providing some of its customers with data comparing their consumption with 100 nearby households in similar-size homes (Ziegler, 2010). The comparison includes a colourful bar chart and 'smiley' faces – two for 'great' and one for 'good'.[5] Such 'positively valenced emoticons' (Schultz et al., 2007, p. 430) are an example of a second type of social norm: the injunctive norm. Whereas descriptive norms affect 'the perception of prevalence,' injunctive norms affect perceptions of approval or disapproval of particular behaviours within a culture (Schultz et al., 2007, p. 429). Programmes that signal approval of conservation behaviours are credited with a 2–3 per cent decrease in energy use (Suzukamo, 2009). Similar results have been observed with regard to natural gas usage (Allcott, 2009; Ayers et al., 2009; Schultz et al., 2007).

Properly identifying the social norms held by target groups is a critical step when crafting messages designed to tap into presumed social norms, a point made by Costa and Kahn (2010a, 2010b) in their work on the influence of ideology in motivating behavioural change. In their randomized field experiment of Californian households receiving a home energy report providing their own electricity usage relative to their neighbours, Costa and Kahn (2010a) find that 'the effectiveness of energy conservation "nudges" depend on an individual's political views' (p. 2). While liberals and environmentalists are already more energy efficient than conservatives, making it more difficult for them to further reduce consumption, these groups are nonetheless more responsive to peer comparisons of home energy use than most people (Costa and Kahn, 2010b). Certain subsets of registered Republicans, on the other hand, actually increased their electricity consumption when provided with peer comparisons (Costa and Kahn, 2010a).

INNOVATION, BEHAVIOUR AND COMMUNITY ENERGY SYSTEMS

All of these various disciplinary perspectives converge to a single, important conclusion, namely, that the successful diffusion of technology, or more generally energy-related programming, depends as much upon the technology itself as the political and social context in which that technology is located (Frankel, 1981). In the case of distributed or alternative energy systems, an effective diffusion process would begin with local agents who possess technical expertise regarding the behavioural changes required of householders or business owners. As pointed out above, however, caution must be exercised since, while experts are credible agents for the dissemination of knowledge, particularly when individuals are faced with the decision of adopting a fairly complex technology such as solar thermal or photovoltaic systems, they are less credible for the purposes of adoption (Rogers, 1983).

An effective diffusion process would therefore combine locally based experts, perhaps provided by local units of government (Ciglar, 1981), with a network of 'early adopting' peers capable of providing the psychological assurance so critical to the adoption decision. In order to respect what Rogers (1983) refers to as the principle of 'homophily', these peers would ideally share attributes similar to the late adopters. 'Diffusion vehicles' of this sort would also align with the type of distributed generation technology generally used in community energy initiatives, that is, community wind, rooftop solar thermal or photovoltaic systems, and/or conservation and efficiency programmes, as well as the development of informal socializing or 'neighbouring' activity critical for the type of community engagement that could persuade people to adopt new energy technology or practices (Berkowitz, 1996; Unger and Wandersman, 1985).[6]

Unfortunately, there are precious few 'community-based energy organizations' that combine a degree of expertise with 'neighbourliness'. Nor are there very many programmes that effectively combine supportive institutional work with broad-based bottom-up initiatives that engage a significant number of community residents. Instead, community energy has largely been limited to either 'top-down' institutional options or community initiatives based upon 'weak' governance structures such as landowner-based utility-scale wind projects (Hoffman and High-Pippert, 2005a, 2010).

The failure to link the two sides of community energy was evidenced in two recent reviews of community energy initiatives in the American state of Minnesota.[7] The first of the reviews examined institutional efforts regarding long-term energy consumption goals.[8] Several findings stood out. First, there were few, if any, well-articulated long-term energy-efficiency or emission reduction goals. Second, those that did have such goals in place directed their efforts largely at various internal activities, such as reducing carbon dioxide emissions associated with municipal operations, increasing renewable electricity in municipal operations and improving the energy efficiency of public buildings.

Third, the connection between institutional and citizen-led efforts was marginal and generally limited to the use of low-cost communication outreach vehicles such as city-sponsored newsletters, websites and blogs. Messages were simple and generic, that is, encouraging residents to purchase energy-efficient appliances, offering energy hints and tips such as how to 'winterize' windows, advice on obtaining tax benefits associated with energy-efficient appliances or spotlighting 'green' businesses. To do otherwise would, as one official noted, require city staff to expend considerable 'resources and time' that, in an era of retrenchment and government downsizing, are unlikely to be available. As one official noted, '[W]e have been just trying to survive . . . With local government aid cuts, we've lost upwards of $2 million from a $15 million [budget]. That's huge.' Officials also acknowledged that developing meaningful ties with citizen-led initiatives aimed at inducing changes in the behaviour of households and firms would require investments in 'social networking and cooperation', which as one official said, 'isn't particularly easy to do . . . We need people to take advantage of the programs. They've got to do something extra to save energy and that is the difficult part. They might do that for a while, but it probably won't last.'

One exception to this finding was the largest city in the survey, which does, in fact, have numerous programmes in place that try to engage with community-based, bottom-up initiatives and which are built upon the principles found in both the innovation/diffusion and social norms literatures. The city's climate change micro-grant programme, for instance, provides grants of between $500 to $10 000 to neighbourhood groups, places of worship, not-for-profit organizations, and various other organizations. Most important, these grants are 'all social marketing based, so it's not the government telling you what to do, it's peers [and] groups that don't have energy conservation as their first priority'. Faith-based organizations, for instance, seldom have an energy-related mission, but according to city officials, when approached, they often see energy-related programming as 'a way of taking care of God's creation. That way [the city] can expand our universe a bit more.'

If local institutions seldom target citizen-based groups in the energy conservation efforts, citizen-based groups are often equally wary of local institutions, an attitude evidenced by a recent study of Minnesota's Clean

Energy Resource Team (CERTs) programme.[9] Between 2006 and 2009, CERTs has awarded 42 community grants of between $1000 and $7000. Grant recipients included the variety of organizational types specified in Figure 12.1, that is, newly formed citizen groups with no professional staff that were created expressly for the purpose of developing an energy-related initiative, professionally led groups focused on an energy-related activity that arise out of existing 'community assets', such as faith communities and scout troops, and local institutions such as schools that serve as a natural meeting ground for particular members whose interests may coalesce around an energy-related initiative (Verba et al., 1995).

Programme objectives specified by the grantees were equally diverse. In some cases, the funds were meant to assist in the installation of specific supply-side technologies such as small-scale wind or solar demonstration projects or even more exotic applications such as off-grid cold storage. Other initiatives focused on demand-side strategies, with recipients intending to develop material suitable for distribution to their neighbours, students in local schools or the community at large. Despite the diversity of organizational forms and the scope of work, community-based initiatives are remarkably consistent in their failure to collaborate with local units of government. Thus, of the 17 community-based programmes funded by CERTs, only one identified cooperation with a local authority as a key element of the proposed work plan, in this case, a local campus of the state university system.

Clearly, the gap between the 'community as local authority' and 'community as citizen' is wide; it is also, at least to some extent, easy to understand. From the perspective of the city official, for instance, community organizations can be unruly and difficult to manage, particularly if they are led by a charismatic leader who may or may not be able to sustain his or her effort over time. The same is true for rank-and-file volunteers, who often experience fatigue and a dwindling ability to commit time and energy to 'the cause' over the generally long period of time required to bring a project to fruition. Citizen-led organizations may also suffer from a lack of clear direction or may be labouring under an ambiguous set of objectives. Conversely, community groups, particularly those with no professional staff, may be inspired by what city officials perceive to be narrow and unrealistic objectives. Under such circumstances, investing increasingly scarce resources of time and money in such organizations is hardly an appealing prospect. The problem is only compounded when outcomes are as difficult to quantify as emissions reductions or energy savings on a city-wide basis.

By the same token, leaders of community organizations may not want to become involved with, or, as they see it, subject to, the dictates of a local authority. Fearing a takeover of who they are or what they represent, a neighbourhood organization, a small not-for-profit organization or even that group of parents being hectored by their children to 'do something for the planet' may well prefer to limit their work to the few blocks they know and understand and feel capable of influencing. Going beyond their geographically or socially familiar boundaries or having to work with planners, elected officials or even unfamiliar community members may well exceed the reservoir of resources available to those with full-time jobs, parental responsibilities, lawns to mow or gardens to tend.

CIVIC ENGAGEMENT AND THE ELECTRICITY SYSTEM

Overcoming these sorts of barriers is of critical importance if the principal lesson of the diffusion of innovation and social norms literatures is to be realized, namely, that programmes based on peer-to-peer and neighbour-to-neighbour relations provide the most effective means of motivating the behavioural change required under any reasonably effective emissions reduction programme. However, both local officials and civically engaged citizens seeking to broaden the base of democratic participation face at least two unfortunate facts. First, even in places with high levels of civic participation, the best designed programmes can still only engage a vanishingly small percentage of a locality's population. Second, it may well be the case that, despite the best hopes of democratic theorists, the vast majority of people prefer the possibilities rather than the actual practices of democratic participation (Hibbing and Theiss-Morse, 2002). Given these circumstances, if people are really going to become active participants in the design of an energy system that is truly sustainable, that is, one that will stand a decent chance of inducing the behavioural changes needed to significantly reduce climate-altering emissions, then a way must be found to engage a significantly larger number of people than is currently the case.

There is, of course, a substantial literature regarding the nature and practice of civic engagement. 'Service' and 'volunteerism', for instance, are often held out as authentic forms of civic engagement and democratic participation (Levine, 2011). Many universities, for instance, have developed so-called service learning programmes, which proponents claim lead to robust sets of positive outcomes, that is, helping young people to thrive by enlisting them as contributors to society, allowing them to gain an appreciation of social diversity and, particularly in the case of lower-income and marginalized populations, getting them 'in the habit of speaking, advocating and voting within the system and thereby helping to reform it' (Levine, 2011). Even advocates of service, however, admit that these outcomes are more suspected than proven. Other critics point out that service programmes 'tend to encourage a distinction between the active server and the passive recipient . . . marginalize civic engagement as something to be done temporarily and unprofessionally, not as an aspect of one's life work, and . . . ignores questions of power' (Levine, 2011).

A somewhat higher bar for engagement is set by a variety of programmes that require participants to scrutinize and deliberate on various policy options being considered by government authorities. These include Fishkin's (1995) deliberative polling process, citizens juries (Hoffman and Matisone, 1997), Denmark's technology consensus conferences and Oregon's watershed councils. The latter is described by Prugh et al. (2000, p. 148) as an effort to involve a broad base of 'stakeholders directly in the resolution of disputes that affect their homes and communities, thereby building trust and reinforcing local commitment'. Participatory processes for urban settings, often at a neighbourhood level, have also been described by Berry et al. (1993).

Much like traditional notions of advocacy and mobilization, however, none of these processes require a sustained level of engagement over time. Instead, individuals are asked to deliberate or simply respond to a specific issue for a limited amount of time and often with a group of strangers rather than with neighbours or other community members with whom they might develop sustained, long-term relationships. As a result, these efforts do not typically generate the sort of social capital or civil society organizations required to fundamentally alter the behavioural changes required for more sustainable energy systems.

One approach that has the potential to overcome these problems is the increasingly popular notion of 'public work' that Boyte (2011) defines as 'self-organized efforts by a mix of people who create things, material or symbolic, of lasting civic value' (p. 94). Boyte (2011, p. 94) traces the origins of public work to:

> [C]ommunal labor traditions and practices around the world associated with the commons, with recurring themes of importance for larger system change today. These include egalitarian and cooperative effort across divisions; practical concerns for creating shared collective resources of a community; adaptability; and incentives based on appeal to immediate interests combined with cultivation of concern for long term community well-being.

What would a community energy system driven by the notion of public work look like? First, a public work-driven reconsideration of the electricity system requires that governance remains in the hands of community members, that is, it would demonstrate the sort of 'strong governance' noted in Figure 12.1. Given the inherent complexity of the electrical system, this is a particularly difficult objective. As any well-meaning 'citizen-activist' soon finds out, trying to come to grips with the workings of the electricity system means confrontation with a blizzard of complex technical information, daunting and difficult to understand regulatory systems and a mountain of acronyms (Hoffman and High-Pippert, 2005b). As a result, 'public work' can quickly become a full-time, and fatiguing, job. Yet, while the sheer complexity of the electrical system requires that expertise be part of the process, extreme care must be taken to avoid violating the central premises of both the innovation literature, that is, successful diffusion depends upon the degree of affinity between early adopters and the target population, and the primary principle of public work, that is, community governance must remain strong. Campaigns and processes designed, administered and implemented by institutionally sponsored experts, even if they do not come equipped with white lab coats, seldom respect either of these demands.

A second important feature of public work is an understanding that individuals and communities have interests that are both practical and principled. As Boyte (2011) points out, 'public work avoids exhortations about what professionals . . . or institutions should do. Rather, public work connects individual and institutional self-interests to citizenship and the public good' (p. 95). Clearly, a more sensible, sustainable electricity system falls squarely within this orbit, offering, as it very well might, both substantial private benefits, such as lower energy costs for both households and firms, and immense social benefits, that is, reductions in sea level rise, less violent and spasmodic weather patterns, and myriad other effects associated with global climate change. A community energy system directed by a public work process would elaborate these interests and seek to discover pathways most likely to satisfy them.

Fortunately, the means for realizing this process of discovery is embedded within a public work process, the central goal of which is to 'liberate energies by basing organizing efforts on questions' (Boyte, 2011, p. 95.) As indicated by the previous discussion regarding the concept of community energy, there is no shortage of questions to be addressed, beginning with the most fundamental: should a community energy system be built on fossil fuels and/or renewable energy sources or on aggressive demand-side programmes meant to reduce household and/or firm consumption? From there, those engaged in a public work process can begin to address the questions of programme design, many of which are centered around the problem of behavioural change: what are the best and most effective means of motivating their neighbours, colleagues and acquaintances to change their energy consumption decisions? How and under what circumstances can local institutions and associations incorporate an 'energy focus' into their activities? What sorts of messages might appeal to the average church or synagogue-goer? How might a local scout troop work with other community organizations to bring about changes that will result in energy savings for both individual households and the organizations themselves? What sort of curriculum will most effectively encourage students, teachers and staff to turn off the lights, thereby reducing energy consumption in schools and saving taxpayer money?

Finally, an electricity system informed in its design by the principles of public work is not about designing the next great technical innovation or determining the best way to engineer a nuclear power plant or even a utility-scale wind machine. It is instead, and at its most basic, about forging the sorts of public relationships necessary to realize the transition to a sustainable electricity system. Again, as Boyte (2011) notes, 'seeing institutions as communities, building public relationships, undertaking intentional changes in their cultures to make them more public, and thinking in political terms' is the distinguishing feature of public work (p. 95). While such relationships no doubt begin with individuals, whether they are neighbours and/or fellow business owners, the core of public work must revolve around communities of people and the variety of institutional actors that populate any locality's landscape.

CONCLUSION

So long as institutional authorities such as cities continue to treat their residents as atomistic actors, unconnected to their neighbours, friends and peers, community energy will remain a largely untapped resource for change. City-sponsored newsletters attesting to a few modest steps taken at or by 'city hall' might well provide some useful information to residents concerned about how the city is spending their tax dollars. And home energy reports accompanying once-a-month utility bills no doubt provide homeowners with a positive incentive for reducing energy consumption. Neither, however, harnesses the power of the actual 'community' within community energy.

At the same time, community organizations must be persuaded that local authorities can, in fact, be something other than an obstacle in the pursuit of their goals. Despite increasingly tough financial realities, cities, counties, municipal utilities and so on possess significant resources that can be effectively leveraged by community residents seeking to influence how their neighbours behave with respect to the consumption of energy and energy services. Perhaps even more importantly, local government institutions have the capacity to institutionalize desirable behaviour through a variety of mechanisms, including building codes, comprehensive planning processes and development requirements. Indeed, for those looking to create a truly sensible energy future built upon locally produced and consumed energy resources, innovations developed by committed citizens working with their neighbours and community peers and implemented by forward-looking local government authorities is likely to be the surest path to success.

NOTES

1. With some 2009 'munis' and 875 cooperative utilities in the USA together supplying electricity to approximately 25 per cent of the nation's electricity users, these are relatively common entities (EIA, 2011).
2. The reasons for failure are numerous though they tend to centre on costs. According to Silverstein, 'things like distribution systems, generation assets, stranded costs, engineering expenses, fuel costs and reliability issues all factor in the ultimate price. In the end, the anticipated price tag is always more than expected' (2011).
3. Behavioural changes can encompass a range of demand-side activities, including some simple measures such as changing light bulbs or investing in energy-efficient appliances to more intensive efforts such as the installation of 'smart grid' meters that would allow customers access to response demand opportunities (Cappers et al., 2011). Supply-side measures might include the installation of rooftop solar photovoltaic panels or solar thermal systems.
4. One exception to this general trend is Rincke's (2006) study of innovations on the part of American school districts.
5. Xcel customers who do not fare well in comparison to their neighbours do not receive frowns, rather, they are told that they used more than average (Ziegler 2010). Frowns, apparently, are resented to the extent that they have been known to actually cause people to increase their consumption.
6. This notion is also familiar to political scientists, who use the language of 'networks' and 'environment' to examine the same processes and their impact on civic engagement and 'the social flow of political information' necessary for the long-term success of community energy initiatives (Green and Brock, 2005; Huckfeldt and Sprague, 1987).
7. Minnesota is a northern tier, upper Midwestern state that has long been recognized for its so-called 'moralistic' political culture, that is, one characterized by high levels of citizen participation in civic activities ranging from voter turnout to volunteering rates (Elazar, 1999; McDonald, 2008; Scott, 2009). Citizens of Minnesota also rank above the national average on measures of 'participation in a community project' (25 per cent) and 'involvement in a public discussion of issues' (27 per cent) (Boyte and Skelton, 2009).
8. For this study, interviews were conducted with local government administrators from Minnesota communities working towards meeting and exceeding the state goals for energy efficiency and global warming emissions reductions, as identified by a recent state legislative report (Nelson, 2009). City administrators were initially contacted through an email describing the nature of the research project and requesting an interview; this was followed by a telephone interview with the responsible city official. The cities were located throughout the state and ranged in size from the state's largest city, Minneapolis (population 372811) to the very small outstate city of Mountain Lake (population 2000).

9. The CERTs programme is a collaborative involving the Minnesota Department of Commerce, the University of Minnesota's Regional Sustainable Development Partnerships programme, Rural Minnesota Energy Task Force, the Metro County Energy Task Force, and the Minnesota Project, a non-governmental organization that works on agricultural issues. CERTs teams have been created for six regions in the state, with each team bringing together people from various cities and counties, farmers and other landowners, industry, utilities, colleges, universities and local governments (Hoffman and High-Pippert, 2005a).

REFERENCES

Allcott, H. (2009), 'Social norms and energy conservation', Massachusetts Institute of Technology Center for Energy and Environmental Policy, working paper no. 09–014, Cambridge, Massachusetts.

Asch, S.E. (1951), 'Studies of independence and conformity: a minority of one against a unanimous majority', *Psychological Monographs*, **70**, 1–70.

Ayers, I., S. Raseman and A. Shih (2009), 'Evidence from two large field experiments that peer comparison feedback can reduce residential energy usage', National Bureau of Economic Research working paper No. 15386, Washington, DC.

Baumgartner, F.R. and B.D. Jones (1993), *Agendas and Instability in American Politics*, Chicago, IL: University of Chicago Press.

Berkowitz, B. (1996), 'Personal and community sustainability', *American Journal of Community Psychology*, **24**, 441–59.

Berry, J.M., K. Portney and K. Thomson (1993), *The Rebirth of Urban Democracy*, Washington, DC: Brookings Institution.

Billman, L. (2009), 'Rebuilding Greensburg, Kansas as a model green community: a case study. NREL's technical assistance to Greensburg June 2007–May 2009', available at www.nrel.gov/docs/fy10osti/45135–1.pdf (accessed 17 December 2012).

Boushey, G. (2004), 'The diffusion of policy innovations through citizen initiatives: advocacy networks, policy frames and the transmission of ideas', paper presented at the annual meeting of the Midwest Political Science Association, Chicago, Illinois.

Bowers, C. (2011), 'Greensburg rebuilds four years after tragic tornado', *CBS Evening News*, 27 May, available at www.cbsnews.com/stories/2011/05/27/eveningnews/main20066961.shtml (accessed 17 December 2012).

Boyte, H.C. (2011), 'Public work and the politics of the commons', *The Good Society*, **20** (1), 84–101.

Boyte, H.C. and N. Skelton (2009), *2009 Minnesota Civic Health Index: Integrating Civic Engagement into Community Life*, Minneapolis, MN: Center for Democracy and Citizenship, Augsburg College.

California Energy Commission (2001), *Distributed Generation Strategy*, available at www.energy.ca.gov (accessed 17 December 2012).

Cappers, P., A. Mills, C. Goldman, R. Wiser and J.H. Eto Mass (2011), *Market Demand Response and Variable Generation Integration Issues: A Scoping Study*, October, Berkeley, CA: Ernest Orlando Lawrence Berkeley National Laboratory, Environmental Energy Technologies Division.

Ciglar, B. (1981), 'Organizing for local energy management: early lessons', *Public Administration Review*, **41**, 470–9.

City of Boulder, Colorado (August 2011), *What is Municipalization?*, available at www.bouldercolorado.gov/index.php?option=com_content&view=article&id=14247&Itemid=4635 (accessed 17 December 2012).
Coleman, J.S. (1959), 'Relational analysis: the study of social organizations with survey methods', *Human Organization*, **17**, 28–36.
Costa, D.L. and M.E. Kahn (2010a), 'Energy conservation "nudges" and environmental ideology: evidence from a randomized residential electricity field experiment', National Bureau of Economic Research working paper no. 15939, Washington, DC.
Costa, D.L. and M.E. Kahn (2010b), 'Why has California's residential electricity consumption been so flat since the 1980s? A microeconometric approach', National Bureau of Economic Research, working paper no. 15978, Washington, DC.
Drazin, R. and C.B. Schoonhoven (1996), 'Community, population and organizational effects on innovation', *Academy of Management Journal*, **39**, 1065–83.
Elazar, D.J. (1999), 'Minnesota: the epitome of the moralistic political culture', in D.J. Elazar, V. Gray and W. Spano (eds), *Minnesota Government and Politics*, Lincoln, NE: University of Nebraska Press, pp. 19–49.
Emmert, C.F. and C.A. Traut (2003), 'Bans on executing the mentally retarded: an event history analysis of state policy adoption', *State and Local Government Review*, **35**, 112–22.
(EIA) Energy Information Administration (2011), *Electric Power Industry Overview 2007*, available at www.eia.gov/cneaf/electricity/page/prim2/toc2.html (accessed 19 December 2012).
Fishkin, J. (1995), *The Voice of the People: Public Opinion and Democracy*, New York: Yale University Press.
Frankel, E. (1981), 'Energy and social change: an historian's perspective', *Policy Sciences*, **14**, 59–73.
Galluzzo, T. and D. Osterberg (2006), *Wind Power and Iowa Schools*, Iowa City, IA: The Iowa Policy Project.
Gerber, A.S. and D.P. Green (2000), 'The effects of canvassing, telephone calls, and direct mail on voter turnout: a field experiment', *American Political Science Review*, **94**, 653–63.
Gladwell, M. (2000), *The Tipping Point*, New York: Little, Brown and Company.
Goldstein, N.J., R.B. Cialdini and V. Griskevicius (2008), 'A room with a viewpoint: using social norms to motivate environmental conservation in hotels', *Journal of Consumer Research*, **35**, 472–82.
Granovetter, M. (1974), 'The strength of weak ties', *American Journal of Sociology*, **78**, 1360–80.
Gray, V. (1973), 'Innovation in the states: a diffusion study', *American Political Science Review*, **67**, 1174–85.
Green, M.C. and T.C. Brock (2005), 'Organizational membership versus informal interaction: contributions to skills and perceptions that build social capital', *Political Psychology*, **26**, 1–25.
Haider-Markel, D. (2001), 'Policy diffusion as a geographical expansion of the scope of political conflict: same-sex marriage in the 1990s', *State Politics and Policy Quarterly*, **1**, 5–26.
Hibbing, J. and E. Theiss-Morse (2002), *Stealth Democracy: Americans' Beliefs about How Government Should Work*, New York: Cambridge University Press.
Hoffman, S.M. and A. High-Pippert (2005a), 'Community energy: a social archi-

tecture for an alternative energy future', *Bulletin of Science, Technology and Society*, **25**, 387–401.

Hoffman, S.M. and A. High-Pippert (2005b), 'The power of words: the rhetoric of community and the reality of community energy,' paper presented at the Australasian Political Science Association Conference, 27 September Dunedin, New Zealand.

Hoffman, S.M. and A. High-Pippert (2007), 'Beyond the rhetoric: distributed technologies and political engagement', *in Proceedings of the 7th International Summer Academy on Technology Studies, Transforming the Energy System: The Role of Institutions, Interests and Ideas*, Graz, 28 August, Austria: The Inter-University Research Centre for Technology, Work and Culture (IFZ).

Hoffman, S.M. and A. High-Pippert (2010), 'From private lives to collective action: recruitment and participation incentives for a community energy program', *Energy Policy, Special Issue: Low Carbon Communities*, **38** (12), 7567–74.

Hoffman, S.M. and S. Matisone (1997), *A Citizens Jury on Minnesota's Electricity Future*, Minneapolis, MN: The Jefferson Center for New Democratic Processes.

Huckfeldt, R. and J. Sprague (1987), 'Networks in context: the social flow of political information', *American Political Science Review*, **81**, 1197–216.

Jacobs, J. (1961), *The Death and Life of Great American Cities*, New York: Random House.

Kretzmann, J. and J.L. McKnight (1993), *Building Communities from the Inside Out*, Chicago, IL: ACTA Publications.

Leonard-Barton, D. (1985), 'Experts as negative opinion leaders in the diffusion of a technological innovation', *Journal of Consumer Research*, **11**, 914–26.

Levine, P. (2011), 'What do we know about civic engagement?', *Liberal Education*, **97** 2, available at www.aacu.org/liberaleducation/le-sp11/levine.cfm (accessed 17 December 2012).

Lovins, A. (1977), *Soft Energy Paths: Towards Durable Peace*, New York: Harper and Row.

Markus, M.L. (1987), 'Toward a "critical mass" theory of interactive media, universal access, interdependence and diffusion', *Communication Research*, **14**, 491–511.

McDonald, M. (2008), '2008 General election turnout rates', United States Election Project, available at www.elections.gmu.edu (accessed 19 December 2012).

Mintrom, M. and S. Vergari (1998), 'Policy networks and innovation diffusion: the case of state education reforms', *Journal of Politics*, **60**, 126–48.

Nelson, C. (2009), *Minnesota GreenStep Cities*, Minneapolis, MN: Clean Energy Resource Teams.

Nolan, J.M., P.W. Schultz, R.B. Cialdini, N.J. Goldstein and V. Griskevicius (2008), 'Normative social influence is underdetected', *Personality and Social Psychology Bulletin*, **34**, 913–23.

Norberg-Bohm, V. (1999), 'Stimulating "green" technological innovation: an analysis of alternative policy mechanisms', *Policy Sciences*, **32**, 13–38.

O'Brien, G. (2009), 'Vulnerability and resilience in the European energy system', *Energy and Environment*, **20**, 399–410.

Parks, C.D., L.J. Sanna and S.R. Berel (2001), 'Actions of similar others as inducements to cooperate in social dilemmas', *Personality and Social Psychology Bulletin*, **27**, 345–54.

Peters, M., P. Sinclair, and S. Fudge (2011), 'Promoting sustainable living through community groups and voluntary organisations', RESOLVE working paper series, 08–11, Centre for Environmental Strategy, University of Surrey, Guildford.

Prugh, T., R. Costanza and H. Daly (2000), *The Local Politics of Global Sustainability*, Washington, DC: Island Press.

Rice, R. (2009), 'Diffusion of innovations: integration and media-related extensions of this communication keyword', paper presented at the annual meeting of the International Communication Association, 21 May Chicago, Illinois.

Rincke, J. (2006), 'Policy innovation in local jurisdictions: testing for neighbourhood influence in school choice policies', *Public Choice*, **129**, 189–200.

Rogers, E.M. (1983, 1995), *Diffusion of Innovations*, New York: The Free Press.

Rosen, E. (2000), *The Anatomy of Buzz: How to Create Word-of-Mouth Marketing*, New York: Random House.

Schultz, W.P., J.M. Nolan, R.B. Cialdini, N.J. Goldstein and V. Griskevicius (2007), 'The constructive, destructive, and reconstructive power of social norms', *Psychological Science*, **18**, 429–34.

Scott, S. (2009), *In Tough Times, Volunteering in America Remains Strong*, Washington, DC: Corporation for National and Community Service.

Sherif, M. (1936), *The Psychology of Social Norms*, New York: Harper.

Silverstein, K. (2011), 'Boulder's bold move to condemn Xcel', 11 August available at www.energybiz.com/move-condemn-xcelarticle/11/08/boulders-bold (accessed 19 June 2013).

Suzukamo, L.B. (2009), 'Thy neighbour's energy usage', *St. Paul Pioneer Press*, 26 July D1.

UK Low Carbon Transition Plan: National Strategy for Climate and Energy (2009), London: Department for Energy and Climate Change.

Unger, D.G. and A. Wandersman (1985), 'The importance of neighbours: the social, cognitive, and affective components of neighbouring', *American Journal of Community Psychology*, **13**, 139–69.

Valente, T.W. (2005), 'Network models and methods for studying the diffusion of innovations', in P. Carrington, J. Scott and S. Wasserman (eds), *Models and Methods in Social Network Analysis*, Cambridge: Cambridge University Press, pp. 98–115.

Verba, S., K.L. Schlozman and H.E. Brady (1995), *Voice and Equality: Civic Voluntarism in American Politics*, Cambridge, MA: Harvard University Press.

Walker, J.L. (1969), 'The diffusion of innovations among the American states', *American Political Science Review*, **63**, 880–99.

Wang, F.Y., K.M. Carley, D. Zeng and W. Mao (2007), 'Social computing: from social informatics to social intelligence,' *IEEE Intelligent Systems*, **22**, 79–83.

Weatherford, M.S. (1982), 'Interpersonal networks and political behaviour', *American Journal of Political Science*, **26**, 117–43.

Ziegler, S. (2010), 'Social pressure to save energy?', *Minneapolis Star-Tribune*, available at www.startribune.com/business/81287582.html?refer=y (accessed 19 December 2012).

Index